SPATIAL PLANNING AND URBAN DEVELOPMENT IN THE NEW EU MEMBER STATES

Spatial Planning and Urban Development in the New EU Member States

From Adjustment to Reinvention

Edited by

UWE ALTROCK
Brandenburg Technical University, Germany

SIMON GÜNTNER
URBACT

SANDRA HUNING
Urban Researcher, Berlin, Germany

DEIKE PETERS
Technical University Berlin, Germany

Routledge
Taylor & Francis Group

LONDON AND NEW YORK

First published 2006 by Ashgate Publishing

2 Park Square, Milton Park, Abingdon, Oxon OX14 4RN
711 Third Avenue, New York, NY 10017, USA

*Routledge is an imprint of the Taylor & Francis Group,
an informa business*

First issued in paperback 2016

British Library Cataloguing in Publication Data
Spatial planning and urban development in the new EU member
    states : from adjustment to reinvention. - (Urban and
    regional planning and development series)
    1. City planning - Europe, Eastern 2. City planning - Europe,
    Central 3. Sustainable development - Europe, Eastern
    4. Sustainable development - Europe, Central 5. City planning -
    Europe, Eastern - Case studies 6. City planning - Europe,
    Central - Case studies 7. Sustainable development - Europe,
    Eastern - Case studies 8. Sustainable development - Europe,
    Central - Case studies
    I. Altrock, Uwe
    307.1'216'0947

Library of Congress Cataloging-in-Publication Data
Spatial planning and urban development in the new EU member states :
    from adjustment to reinvention / edited by Uwe Altrock ... [et al.].
        p. cm. -- (Urban and regional planning and development series)
    Includes index.
    ISBN 0-7546-4684-X
    1. Regional planning--Europe, Central. 2. Regional planning--Europe, Eastern.
    3. City planning--Europe, Central. 4. City planning--Europe, Eastern. 5. Urban policy-
    -Europe, Central. 6. Urban policy--Europe, Eastern. 7. Urbanization--Europe, Central.
    8. Urbanization--Europe, Eastern. 9. European Union. I. Altrock, Uwe. II. Urban and
    regional planning and development.

    HT395.E36S63 2006
    307.1'20943--dc22

                                                                            2005028882

ISBN 978-0-7546-4684-6 (hbk)
ISBN 978-1-138-27321-4 (pbk)

Contents

PART I   EAST-WEST PERSPECTIVES ON SPATIAL PLANNING AND URBAN DEVELOPMENT IN THE ENLARGED EU

PART II   SPATIAL PLANNING AND SUSTAINABLE DEVELOPMENT IN THE NEW EU MEMBER STATES

List of Figures

List of Tables

List of Contributors

Uwe Altrock, Junior Professor of Urban Structures: Assessment and Conservation Policies, Technical University of Cottbus.

Zigmas J. Daunora, Professor of Urban Design, Technical University of Vilnius.

Simin Davoudi, Professor and Director of the Centre for Urban Development and Environmental Management, Leeds Metropolitan University; President, AESOP Planning Association.

Kaliopa Dimitrovska Andrews, Architect; Director of the Urban Planning Institute of the Republic of Slovenia; Professor of Urban Planning, University of Ljubljana.

Zoltán Dövényi, Geographic Research Institute at the Hungarian Academic of Science.

Susanne Frank, Junior Professor of Urban Sociology, Humboldt University, Berlin.

Yaakov Garb, Hebrew University, Jerusalem; Central Eastern Europe & Middle East Regional Coordinator, Institute for Transportation and Development Policy, New York.

Simon Güntner, Lecturer in Urban and Regional Sociology, Technical University of Berlin.

Sandra Huning, Lecturer in Urban and Regional Sociology, Technical University of Berlin.

Jakob Hurrle, Master of Science in Urban Planning, Technical University of Berlin.

Jiřina Jackson, Architect; Institute for Transportation and Development Policy, Prague Office.

Prančiskus Juškevičus, Professor, Technical University of Vilnius.

Zoltán Kovács, Member of the Hungarian Academy of Sciences; Visiting Professor, University of Leipzig.

Klaus R. Kunzmann, Jean Monnet Professor of European Spatial Planning, University of Dortmund; Founding President, AESOP Planning Association.

Piotr Lorens, Assistant Professor, Department of Urban Development, Technical University of Gdansk.

Inara Marana, Land Use Planning Department, City of Riga.

Deike Peters, DFG Post-Doctoral Fellow, Center for Metropolitan Studies, Technical University of Berlin.

Sampo Ruoppila, Sociologist; Ph.D. Student, University of Helsinki.

Richard Sharpley, Professor in Tourism, University of Lincoln (UK), Head of the Department of Tourism and Recreation.

Luděk Sýkora, Assistant Professor for Urban Geography, Charles University, Prague.

Conrad Thake, Architect and Engineer, working on miscellaneous urban planning issues in Malta.

Mareile Walter, Ph.D. Student, Institute for Physical Planning, Technical University Blekinge, Karlskrona, Sweden

Acknowledgements

The present volume has its origins in a special 'Eastern Enlargement' edition of the *Planungsrundschau*, a German-language journal / publication series on planning theory, politics and practice. While the selection and arrangement of the current volume is somewhat different from its German counterpart, we would nevertheless like to thank all authors involved in the bi-lingual endeavour. Commissioning, translating and then editing two similar but nevertheless slightly varying sets of papers in two languages for which only comparatively few of the authors were native speakers turned out to be a rather monumental task, and it was only made possible by the gracious cooperation of everyone involved. The editors would also like to acknowledge the Technical University of Berlin and the Brandenburg Technical University of Cottbus as institutional bases for this project. In Cottbus, Anka Laschewski provided some valuable editorial assistance. We are also indebted to the editors of Ashgate's Urban and Regional Planning and Development Series, Peter Roberts and Graham Haughton, for recommending this set of papers to Ashgate, and to the commissioning editor at Ashgate, Valerie Rose for the excellent cooperation during the preparation of this volume.

Chapter 1

Spatial Planning and Urban Development in the New EU Member States – Between Adjustment and Reinvention

Uwe Altrock, Simon Güntner, Sandra Huning and Deike Peters

Introduction

The accession of ten new member states to the EU represents a historical milestone for governance and spatial development in Central Eastern Europe. In May 2004, the three Baltic countries Estonia, Latvia and Lithuania, as well as Poland, the Czech Republic, Slovakia, Hungary, Slovenia and the two Mediterranean islands of Malta and Cyprus joined the European Union, thereby profoundly altering the overall institutional map of Europe. As a group, the ten new members are highly heterogeneous, both spatially and socio-economically. Even the former COMECON countries, with their common experience of a rapid transition from centrally-planned to democratic, market-oriented states, have very different regional economic and land use structures, also owing to the diverse political and socio-economic history of the various newly created and reconstituted states. In addition, the transformation processes had different effects on the different societies. In some countries, one can observe a shrinking of the population (Hungary, Czech Republic, Lithuania and Estonia) while others are growing (Poland, Malta and Cyprus).

Settlement patterns in the new member states also vary widely; some countries are highly urbanized, others predominantly rural. There is no large metropolis matching the size of London or Paris, and only three cities have more than a million inhabitants: Budapest, Warsaw and Prague. In some of the countries, there is only one major metropolis which serves all central functions (Tallinn, Budapest, Riga, Prague), in other countries there are several larger cities (Poland, Lithuania, Slovakia). In Slovenia, even the capital Ljubljana only has about 260,000 inhabitants, and Valetta, the capital of Malta, has less than 8,000 inhabitants.

Over the course of the last decade, the larger cities typically lost population, for various reasons: low birth rates, suburbanisation, dramatic rises in rents and property values in the inner cities, economic restructuring and job loss. A particularity in the Baltic States is the emigration of the Russian population. The transition to a market economy and a capitalist society remains the dominant theme in the cities: during socialist times, the cities were the centres of industrialisation. Today they are experiencing massive de-industrialization – a rapid structural chance which they have to master. Consumption and production patterns have been fundamentally rearranged. Western European and North American multinationals fight over future market shares in the region, strategically developing hypermarkets and other big box retail stores on greenfield sites along major access roads, thus increasingly fostering a culture of automobile dependent consumerism. The privatisation of the housing stock is a central planning problem in the cities, as are the transportation and environmental consequences of urban sprawl (see KPMG 2004a).

In some of the transition countries, the social and economic transformations also brought about radical readjustments in the national urban hierarchy and new challenges for regional development. Opposite the new challenges (like, for example, the need for developing future-oriented, comprehensive land use plans restricting the currently more or less uncontrolled urban sprawl at the edges of the major cities) one often finds an idealized conceptualisation of purely market-based instruments. At the same time, the political and planning systems have to account for far-reaching EU regulations and directives. In the new member states, urban issues are hardly addressed in a comprehensive manner at the national level (with Slovenia being a notable exception). Local governments often set their own agendas, but in a setting of chronically under-financed municipal budgets and fragmented administrative structures, the private sector has a significant influence over municipal decision-making. There are thus increasing calls for developing national urban policies in order to coordinate medium to long-term development (see KPMG 2004a). The larger cities in Central and Eastern Europe are also showing a significant interest in an urban policy framework provided by the EU.

The aim of this volume is to present a preliminary review of developments and planned interventions in the ten new member states since the beginning of the transition, seen in light of their recent accession to the EU. In order to present such a picture, one can take one of two possible perspectives. On one hand, one might focus on the social and economic transition as a starting point. This places socio-economic transformations and their effects on urban regions centre stage, and one would then have to look at how public administrations and local planning stakeholders deal with these processes and try to organize them spatially. Another possible point of entry is the planning system. Such a perspective would focus more on the question of how the set of involved actors and the new governance apparatus have changed in relation to previous governance arrangements. However, since those two perspectives are in fact complementary, it is precisely the point of intersection between these two approaches which lies at the heart of our volume. Further, it is necessary to contextualise this dual perspective on urban-regional transformation processes within the historical moment of EU accession in May 2004. Ever since the prospect of accession became real, particularly since the beginning of the accession negotiations in 1997, the process of Europeanisation – the adoption of the 'rules' of the European Union – has overshadowed the transition. Accession into the union of European states was always linked not only to the establishment of a functioning market economy, but also to political conditionalities – constitutionality, democracy, protection of minorities – and the adoption of the European Union's vast *acquis communautaire*. This 'conditionality of Europeanisation' had ambivalent effects: the price of a successful rapid adoption of the rules was a slowdown or even rollback of decentralization processes in some advanced states, where executive forces were (re)invigorated. At the same time, some observers note a tendency towards a 'transposition without implementation,' i.e. a discrepancy between the formal adoption of EU rules and their actual realization (see e.g. Schimmelfennig 2004: 266).

In our discussions with the authors, we have particularly tried to address the question of concrete changes as a direct result of the accession. While it is still too early to provide a comprehensive answer to this question just one year after the fact, thus far, it seems apparent that the most crucial changes already took place in the preliminary stages. Day-to-day political and administrative operations in the respective states have been dominated by the prospect of accession and the adjustment to existing EU law due to the impending necessary future adoption of the entire *acquis communautaire*. The actual date of accession itself was therefore not a significant factor for ongoing spatial developments and the related planning interventions. It is only now that the accession is official that the EU's regional and environmental policy framework can provide the respective cities and regions with a welcome 'structure of possibility' in order to better manage the massive transformation, even though the pre-accession funds and instruments already played a key role in this regard (comp. European Commission 2004: 170ff).

Given the highly heterogeneous situation, it makes little sense to speak of 'one Eastern European (planning-)family' (Malta and Cyprus are exempt here anyway). Depending on the national context and the particular path of reform chosen, planning has a very different meaning in the various countries. Additionally, there are vast differences in the degree of fiscal and administrative decentralisation as well as in geographical size, meaning that the administrative and decision-making structures of the individual countries are often very hard to compare, especially since they are at the same time still undergoing constant modification within those countries. This is why we consider it more appropriate to emphasize the different paths of transformations rather than offer a premature typology. Pre-soviet legacies (such as resurrected traditions and historical (transport-)connections) are being considered as one factor among others here.

The complex interrelationship of enabling and disabling factors for the development of new planning systems – as well as the different size, position, economic base, and political tradition of the formerly socialist Central European countries – points to a significant differentiation of their political and planning future, even when the spatial challenges regarding the transition to a market economy did include a variety of similar features, such as the problem of restitution, the development of self-reliant local governments, the divestiture of the socialist production units (which also effectively ended the dominant influence of socialist industrial policy on settlement systems), the increasing spatial and social differentiation, the privatisation of the housing market etc.

Table 1.1 The population of the EU member states in comparative perspective (in 1,000)

	1995	1999	2003
EU-25	446,808.1	450,677.5	454,552.3
EU-15	371,605.4	375,719.5	380,351.4
Euro zone	299,073.1	302,160.5	306,698.2
Austria	7,943.5	7,982.5	8,067.3
Belgium	10,130.6	10,213.8	10,355,8
Cyprus	645.4	682.9	715.1
Czech Republic	10,333.2	10,289.6	10,203.3
Denmark	5,215.7	5,313.6	5,383.5
Estonia	1,448.1	1,379.2	1,356
Finland	5,098.8	5,159.6	5,206.3
France	57,752.5	58,496.6	59,630.1
Germany	81,538.6	82,037	82,536.7
Greece	10,595.1	10,861.4	11,018.4
Hungary	10,336.7	10,253.4	10,142.4
Ireland	3,597.6	3,734.9	3,963.6
Italy	57,268.6	57,612.6	57,321.0
Latvia	2,500.6	2,399.2	2,331.5
Lithuania	3643	3,536.4	3,462.6
Luxemburg	405.7	427.4	448.3
Malta	369.5	378.5	397.3
Netherlands	15,424.1	15,760.2	16,192.6
Poland	38,580.6	38,667	38,218.5
Portugal	10,012.8	10,150.1	10,407.5
Slovakia	5,356.2	5,393.4	5,379.2
Slovenia	1,989.5	1,978.3	1,995
Spain	39,305.4	39,724.4	41,550.6
Sweden	8,816.4	8,854.3	8,940.8
United Kingdom	58,500.2	59,391.1	59,328.9
Iceland	267	275.7	288.5
Liechtenstein	30.6	32	33.9
Norway	4,348.4	4,445.3	4,552.3
Canada	29,437	N/A	N/A
Japan	125,570	126,056.8	127,273.8 (2004)
USA	261,687	271,626	291,685.1 (2004)

Source: *Eurostat* 2005

So far, analyses of the planning systems in the formerly socialist countries primarily concentrated on Poland, Hungary and the Czech Republic (comp. Newman and Thornley 1996: 35–38, 69–71, but also the numerous special issues in academic journals on Central Eastern European countries.) These publications tended to generalize the reform experiences and paths in Central Eastern Europe. But, as the contributions in this volume show, the reality is much more complex, even if the basic insights regarding the difficulties of the political transition in the 1990s are generally confirmed by these accounts. And it was already obvious by the early 1990s that in the larger countries, and in Poland in particular, one would have to regionally differentiate between the 'transformation winners' in the Western part and the 'transformation losers' in the Eastern part.

Without targeted structural assistance programs, there is a danger that those countries with very dominant capitals located close to the old EU will experience an increasing division into a booming West and a poor East. This polarisation is being exacerbated by the additional contrast between urbanised and rural-agrarian lifestyles. Compared to the old EU, the new member states are still much more characterised by non-urban structures. The example of the Baltic states, however, shows that a geographically peripheral location within the European settlement area as a whole need not necessarily play a decisive role for the prospects for a successful transformation of the country as a whole.

Outline of the Book

In order to adequately deal with regionally- and country-specific dynamics, the following contributions were not devised as a comprehensive list of comparative country studies. Instead, we worked together with the authors in order to identify typical planning and development problems for each country and to then provide an in-depth coverage of the selected focus. So although every accession country is featured in at least one article (the exceptions being Poland and the Czech Republic which have two contributions each), some of the articles concentrate on individual cities while others instead describe the conversion to and the search for appropriate new planning systems at the regional, national and supra-national level. Overall, the book is arranged into three main thematic blocks: Following this introduction, part one opens with three overview pieces that look at *spatial planning and urban development from an EU-wide perspective*. In part two, contributions on Slovenia, Lithuania, Poland, the Czech Republic, Slovakia, Hungary, and Cyprus place issues of *spatial development and sustainable development* at the centre of attention. In part three, case studies from Estonia (Tallinn), Latvia (Riga), Malta (Valetta), Poland (Warsaw), and the Czech Republic then take a closer look at various *special issues in urban planning* such as housing, historic preservation, urban regeneration, commercial/retail development and brownfield reuse.

East-West Perspectives on Spatial Planning and Urban Development in the Enlarged EU

In his contribution on the future East-West agenda of Europe, Klaus Kunzmann outlines six issues of particular relevance to the situation in the new member states which will have to be addressed by politicians and policy makers: growing social and spatial polarisation, structural change and industrial development, agricultural divergence, insufficient transport infrastructure, natural heritage conservation and the brain drain of the qualified labour force. He also makes particular reference to the recent crisis of the EU brought about by the negative referenda in France and the Netherlands. In the remainder of his contribution, he then discusses the future role and a possible revised outline for the European Spatial Development Perspective.

Simin Davoudi identifies three key challenges for spatial planning in the EU accession states. She notes great regional disparities both within the states as well as between them due to their different starting points and developments since the end of the Soviet Regime 15 years ago. Her second point concerns the relationship between economic growth and environment protection as a major field of strategic spatial planning. Finally, she points to the nature and the quality of the institutional context in the emerging regional governance patterns as an important issue.

Susanne Frank sketches a 'short history of European urban politics'. She shows how, over the last two decades, the EU changed the nature of its political interventions, starting with a focus on the environment in the 1980s, followed by a focus on cohesion policies in the 1990s and, most recently, a promotion of 'competitive' cities or regions. The EU's vision of a 'European city' changed from the centre of European civilisation and democracy to the place where 'social market economy' is shaped. Whether, and in which form, urban issues will continue to play a significant role at the level of the EU in a context where cohesion and competition dominate political thinking remains an open question.

Spatial Planning and Sustainable Development

The main aim of the chapter by Kaliopa Dimitrovska Andrews is to give an overview of Slovenian planning and building practice, emphasising the impacts of globalisation and Europianisation processes. The article summarises the individual aspects of the connections between global trends and local urban problems, and attempts to define the role of the vision of sustainable spatial development in Slovenia in orienting local transformations, especially of urban areas. In the introduction, some general facts on Slovenia are presented, followed by a discussion on the influence of Europianisation on the spatial planning and urban policies in member states and consequently Slovenia. In the second part, a brief introduction of the planning system in Slovenia is given and national urban policies are presented. The third part includes an overview of development trends and urban problems at the local level and possible approaches for solving them. Slovenian sustainable development perspectives are discussed in the concluding part of the article.

Principles of regional development consolidated in the first Lithuanian territorial planning scheme described by Zigmas J. Daunora and Prančiskus Juškevičius had a marked influence on the sustainable evolution of the urban system in the intensive urbanization period. With a transition to market conditions, the polarisation in the quality of life in regions has been expanding fast. In the context of the strengthening economy, there is a need for a more precise determination of state regional policy and an escape from speculative ideas in addressing urban development towards concentration. This chapter addresses the need for the establishment of a legal basis for the regulation of city and regional planning and development.

Piotr Lorens discusses various factors influencing the current urbanization tendencies in Poland. Starting from remarks concerning the history of urban development, he brings together different factors shaping the Polish model of urbanization, like land ownership pattern, planning situation, administrative structure, and level of government involvement in the process. The article also describes the effects of introducing the free market without the proper preparation of the urbanization process by the municipalities. It also includes a few remarks on the emerging trends in urbanization – like urban regeneration and gentrification – as well as the development of new types of urban programs.

The chapter by Luděk Sýkora presents an overview of the growth and decline of cities and changes in their internal urban spatial structure in the Czech Republic after 1989. Special attention is given to the most pressing urban problems, namely the formation of inner city brownfields, the decline of housing estates and the negative consequences of sprawl-like suburbanisation. The second part is devoted to the urban policies and planning at the city level and to a discussion of national government policies and programmes that influence urban change. In the Czech Republic, the responsibility for urban development rests primarily with city governments, which are in some instances supported by national government programs, such as housing and regional policies or support to FDI.

In his article on Slovakia, Jakob Hurrle focuses on various development strategies for rural Roma settlements in the eastern part of the country. As a marginalised minority, the Roma were historically subjected to varying policies of assimilation and segregation. Towards the final phase of socialism, attempts at assimilation were increasingly being replaced by a policy of state assistance resulting in reduced capacities and incentives for self-reliance among the Roma communities. Current initiatives on the part of the Slovak government, the EU and international organisations combine an activating social policy with physical upgrading of the settlements. The article concludes that these interventions have been mostly unable to do away with the ghetto-like character of the targeted settlements.

Zoltán Dövényi and Zoltán Kovács try to characterise the patterns of town and city development in Hungary. Above all, the chapter investigates new tendencies that have come up as a consequence of the transition process Hungary has gone through. The authors ask whether Hungarian towns and cities follow trends that have been present in Western European cities and if a tendency towards convergence can be observed. It seems interesting to note that they identify an expansion, transformation and differentiation of the system of towns and cities that has its origin in strong political action at the level of the central state.

Richard Sharpley looks at the role of tourism in the economic development of Cyprus. While the island has witnessed an enormous economic boom due to its attractiveness for tourists (mainly from the UK), the negative consequences cannot be overlooked. The national economy depends so heavily on tourism that some sections of the coastline suffer from excessive development. Although the government has been trying to plan for more sustainable development and to integrate the hinterland, its policies are almost without effect. The author concludes by arguing that the future of the relatively expensive tourist destination is uncertain since the diversity of the island's tourist attractions is limited in comparison to other places.

Special Issues in Urban Planning

Sampo Ruoppila analyses the housing situation and housing policies in Tallinn and the changing picture after the city became the capital of the newly independent Estonia. He describes the Estonian politics of privatisation and of restitution of formerly private apartment buildings. The accelerated residential mobility resulting from this development and from limited state intervention into the housing market have reestablished a diverse pattern of neighbourhoods for different incomes originating in pre-Soviet times and scattered throughout the city. These were complemented by newly-built luxury neighbourhoods in the periphery. The observed pattern may be typical for post-socialist cities, but the future must determine whether the renaissance of redistributive policies by the states and the cities will help limit residential segregation as they do in Tallinn after the turn of the century.

The situation in Riga city region is the focus of the chapter by Inara Marana. The new challenges for the capital of Latvia are a result of demographic, economic and social changes and the ongoing urban development since Latvian independence in 1991. The author describes the Latvian planning system and illustrates the situation with current examples of planning in Metro Riga and the historic core.

Conrad Thake presents examples of regeneration efforts in old Valletta, the capital of Malta. For centuries, the island that was conquered by various powers had to put a certain amount of resources into its defensive structures. Nowadays, the remaining fortifications are one important factor for an urban strategy that tries to revitalise abandoned historic buildings and use them for public and sometimes private purposes in a central heritage location that is only very slowly being repopulated. The results are nevertheless striking especially considering that Valletta is a place where strategic public interventions into the land-use pattern do not have a strong tradition.

Mareile Walter comes back to the case of Poland and focuses more closely on the relationship between city and suburban development in Warsaw. The capital of the largest accession country has become an important hub for international companies trying to penetrate the Eastern European markets, and has therefore witnessed an outright boom of investment. The reinvention of a city centre happened in a very particular environment characterised by the attempt of preserving the rebuilt old town but overcoming the legacy of Soviet influence embodied in the Stalinist high-rise cultural palace, whereas the suburban periphery is, as the one of other big Eastern European cities, being besieged by an armada of shopping centres built by Western European companies.

Finally, Yaakov Garb and Jiřina Jackson take another look at the Czech Republic and describe how urban planning issues surrounding the particular problem of brownfields play out in the Central European context. They reflect on the role of a non-profit advocacy organization in generating change. While the issue of brownfields is generally well recognized in Western Europe, the particular circumstances of Central Europe produced a unique nature and scale of the brownfield problem. The interaction of the brownfield problem with EU accession is also discussed, including some suggestions for making EU funding and planning categories more capable of facilitating brownfield reuse.

Conclusions: Towards Adaptation?

With the chapters of this volume, we hope to provide deeper insights into the multiple challenges and trends that can be observed during an increasing Europeanisation of spatial planning and city development. Many of the contributions in this volume show how policy-makers in the accession states still waver between an outright refusal to revitalise the idea of spatial planning as a tool for better land use management and integrated development on the one hand, and an enforced adaptation to EU coordinating, planning and funding mechanisms on the other hand. The mostly negative experiences with centrally-planned economies – which can even be considered one impetus for the peaceful revolutions in the early 1990s – explain the rather liberal approach towards the transition launched after the end of socialism. The stiff wind of globalisation forcing the former socialist states to quickly adapt to a completely new picture further contributed to it: liberalism seemed to be the appropriate answer to the decline of state-controlled manufacturing giants. Cities and towns, having rid themselves from the often strangling party rule were eager to establish local self-governments and to improve their competitiveness. Unfortunately, lacking planning skills and too great a scepticism towards planning on the part of the newly-elected public officials then often led to uncoordinated private investment. At the same time, the decline of the manufacturing base left behind a huge stock of derelict sites. It is true that inward investment and the establishment of modern office complexes have created modern CBDs in some major cities and attractive historic cores have been revitalised by retailers and hotels. But particularly in the retail sector, it is difficult to compete with the rapidly developing suburban developments, especially since land acquisition and site preparation is still much easier and cheaper in greenfield locations. In this context, tools and incentives for the revitalisation of centrally-located brownfield sites are lacking in many places. The rapid transformation of the economic base has very clear and serious consequences for the future spatial structure of Central European cities and towns.

In light of these socio-economic challenges, an integrative understanding of sustainability promoted by the EU calls for ecologically friendly development and social justice. Only a few states seem to go for plans that promote an integrated understanding of ecological sustainability, whereas issues of social justice and cohesion are gradually gaining importance as segregation and polarisation grows in Eastern European cities.

Facing these multiple challenges, the undue scepticism towards spatial planning and urban development strategies in the accession states in the early and mid-1990s obviously hindered a fresh and timely approach towards comprehensive land use management. Cities in the accession states were therefore quickly confronted with negative trends like suburban sprawl and the decay of inner-city brownfields. It was not before the end of the 1990s that the Central East European accession states gradually started to amend their planning laws so as to lay the foundation for new funding schemes and to prepare for EU structural policies. Although the states have made good progress, the funding of urban regeneration or regional planning in metro areas is still underdeveloped. Partnerships for the redevelopment of brownfield sites are still relatively difficult to build.

On the other hand, the EU accession process has initiated substantial adaptation mechanisms. In countries like Poland, the regions were adapted to better suit the EU terminology. In Slovenia, national planning seems to be able to integrate a wide array of policies into the overall goal of sustainable development. Policies against social polarisation are slowly being developed in cities like Tallinn.

Overall, the efforts to prepare for EU integration and to establish a new understanding of planning and development seem to have produced ambivalent results. Sustainability continues to be one of the central challenges for spatial and urban development, and it is one that needs to be addressed both at the national and the supra-national level. In an era of uncertain transition where even economic growth is at stake, it will remain of particular interest for planners and scholars to observe whether the increasingly liberal course of EU politics – which still talks about 'cohesion' and 'social justice' (but in fact increasingly focuses on strengthening already competitive areas rather than provide wide-spread structural assistance to weak ones) – will produce substantial progress in key urban issues like economic and spatial segregation, inner city revitalisation, limitation of sprawl, social cohesion and ecological modernisation.

Further Reading

European Planning Systems and Policies in General.

Bennett, R.J. (ed.) (1993), *Local Government in the New Europe*. London: Belhaven Press.

Elander, I., Gustafsson, M. (1993), 'The re-emergence of local self-government in Central Europe.' In: *European Journal of Political Research* 23.

European Council of Town Planners (2001), 'Proceedings of the Symposium Spatial Planning in Accession Countries' & 'Implications of EU enlargement for spatial planning in Warsaw,' 18 May 2001.

KPMG (2004a), *National Urban Policies in the European Union*, Amstelveen.

KPMG (2004b), *Cities in the New EU Countries – Position, Problems, Policies*, http://www.minbzk.nl/contents/pages/10238/citiessummary0110.pdf.

Newman, P., Thornley, A. (1996), *Urban Planning in Europe: International Competition, National Systems and Planning Projects*. London: Routledge.

Petrakos, G., Maier, G., Gorzelak G. (eds) (2000), *Integration and Transition in Europe: The Economic Geography of Interaction*. London: Routledge.

Regulski, J., Kocan, W. (1994), 'From communism towards democracy: local government reform in Poland.' In: Bennett, R.J. (ed.), *Local Government and Market Decentralisation*. Tokio: United Nations University Press.

'The EU Compendium of European Planning Systems and Policies' (1996).

EU Affairs, Regional Affairs

Albrechts, L., Moulaert, F., Roberts, P., Swyngedouw, E. (1998), *Regional Policy at the Crossroads: European Perspectives*. London: Jessica Kingsley Publishers.

Bachtler, J., Downes, R., Gorzelak, G. (2000), *Transition: Cohesion and Regional Policy in Central and Eastern Europe*. Aldershot: Ashgate.

Bachtler, J., Downes, R. (2000), 'The Spatial Coverage of Regional Policy in Central and Eastern Europe.' In: *European Urban and Regional Studies* 7: 2, pp. 159–174.

Bachtler, J., Wishlade, F. (n.d.), 'Searching for Consensus: the Debate on Reforming EU Cohesion Policy.' University of Strathclyde: European Policies Research Paper 55, http://www.eprc.strath.ac.uk/eprc/PDF_files/R55SearchingforConsensus.pdf.

Bachtler, J., Downes, R. (1999), 'Regional Policy in the Transition countries: A Comparative Assessment,' European Briefing, *European Planning Studies* 7: 6.

Balchin, P., Sykora, L., Bull, G. (1999), *Regional Policy and Planning in Europe*. London: Routledge.

Gorzelak, G. (1999), *The Regional Dimension of Transformation in Central Europe*. London. Routledge.

Joeniemi, P. (ed.) (1997), *Neo-nationalism or Regionality: Restructuring of Political Space around the Baltic Rim*. Stockholm: NordRegio.

Mannin, M. (ed.) (1999), *Pushing Back the Boundaries: The European Union and Central and Eastern Europe*. Manchester, Manchester University Press.

Mayhew, A. (1999), *Recreating Europe: The European Union's Policy towards Central and Eastern Europe*. Cambridge, Cambridge University Press.

Pavlínek, P. (2004), 'Regional Development Implications of Foreign Direct Investment in Central Europe.' In: *European Urban and Regional Studies* 11: 1, pp. 47–70.

Schimmelfennig, F., Sedelmeier, U. (2005), *The Europeanization of Central and Eastern Europe*. New York: Cornell University Press.

Sedelmeier, U., Wallace, H. (2000), 'Eastern Enlargement.' In: Wallace, H., Wallace, W. (ed.), *Policy-Making in the European Union.* Oxford, Oxford University Press.

Turnock, D. (2002) 'Cross-border Cooperation: A Major Element in Regional Policy in East Central Europe.' In: *Scottish Geographical Journal,* 118: 1, pp. 19–40.

Williams, R.H. (1996), *European Union Spatial Policy and Planning.* London: Paul Chapman Publishing.

City Politics, Metropolitics, Housing Policy

Andrusz, G. (ed.) (1996), *Cities after socialism: urban and regional change and conflict in post-socialist societies.* Oxford: Blackwell.

Clapham, D., Kintrea, K. (1996), 'The patterns of housing privatization in Eastern Europe.' In: Clapham, D., Hegedus, J., Kintrea, K., Tosics, J., (ed.), *Housing Privatization in Eastern Europe.* Westport, Connecticut, London: Greenwood Press, pp. 169–189.

Fassmann, H., Lichtenberger, E. (1996), 'Transformation in East Central Europe real estate, housing and labour markets. Acta Facultatis Rerum Naturalium Universitatis Comenianae.' *Geographica* No. 37/1996, pp. 16–33.

Kovacs, Z. (1998), 'Transformation of the housing markets in Budapest, Prague and Warsaw.' In: Kovacs, Z., Wiessner, R. (ed.), *Prozesse und Perspektiven der Stadtentwicklung in Ostmitteleuropa.* Münchener Geographische Hefte 76, Passau: L.I.S. Verlag, pp. 245–256.

Rehnicer, R. (1998), 'New challenges for urban planning in Central and Eastern Europe.' In: Kovacs, Z., Wiessner, R. (ed.), *Prozesse und Perspektiven der Stadtentwicklung in Ostmitteleuropa.* Münchener Geographische Hefte 76, Passau: L.I.S. Verlag, pp. 63–71.

Decentralisation, Administrative Reform

Bird, R., Ebel, R., Wallich, C. (1995), *Decentralization of the Socialist State.* Washington: World Bank.

Horváth, T. M. (2000), *Decentralisation. Experiments and Reforms.* Budapest: Open Society Institute, Local Government and Public Service Reform Initiative.

Peteri, G. (2002), *Mastering Decentralisation and Public Administration Reforms in Central and Eastern Europe.* Budapest: Open Society Institute

Regulski, J. (2003), *Local Government Reform in Poland. An Insider's Story.* Budapest: Open Society Institute.

Wetzel, Deborah (2001), *Decentralization in the Transition Economies: Challenges and the Road Ahead.* World Bank Study No. 25132.

Baltic States

ARL (ed.) (2001), *Regional Planning and Development around the Baltic Sea*. 2nd Planners´ Forum in Erkner/Berlin, 4–6 May 2000. BSR INTERREG II C project – The Baltic Manual: sharing expertise in spatial planning. Hannover: ARL.

Kuklinski, A. (ed.) (1995), *Baltic Europe in the Perspective of Global Change*. European Institute for Regional and Local Development, University of Warsaw, Europe 2010 Series, Volume 1, Warsaw, Oficyna Naukowa.

Linkola, T. (1994), *Land reform and territorial planning in Latvia*. Helsinki: Assoc. of Finnish Local Authorities.

Petkevicius, A. (2001), *Regional policy in Lithuania on the eve of EU structural funds*. Vilnius: Open Society Institute.

Vanagas, J. (2002), 'Planning urban systems in Soviet times and in the era of transition: the case of Estonia, Latvia and Lithuania.' In: *Geographia Polonica*, 75: 2, pp. 75–100.

Poland

Blazyca, G., Heffner, K., Helinska-Hughes, E. (2002), 'Poland – Can Regional Policy Meet the Challenge of Regional Problems?' In: *European Urban and Regional Studies* 9: 3, pp. 263–276.

Czarniawska-Joerges, B. (2002), *A city reframed: managing Warsaw in the 1990s*. Amsterdam: Harwood Academic.

Czerny, M. (2002), 'Introduction: Uneven Urban and Regional Development in Poland.' In: *European Urban and Regional Studies* 9: 1, pp. 37–38.

Czerny, M., Czerny, A. (2002), 'The Challenge of Spatial Reorganization in a Peripheral Polish Region.' *European Urban and Regional Studies* 9: 1, pp. 60–72.

Głebocki, B., Rogacki, H. (2002), 'Regions of Growth and Stagnation in Poland: Changes in Agriculture, Industry and International Markets.' *European Urban and Regional Studies* 9: 1, pp. 53–59.

Hardy, J. (2004), 'Rebuilding Local Governance in Post-communist Economies: the Case of Wrocław, Poland.' In: *European Urban and Regional Studies* 11: 4, pp. 303–320.

Lisowski, A., Wilk, W. (2002), 'The Changing Spatial Distribution of Services in Warsaw.' In: *European Urban and Regional Studies* 9: 1, pp. 81–89.

Markowski (ed.) (2002), *Multipolar patterns of urban development: polish perspective*. Warschau: Komitet Przestrzennego Zagospodarowania Kraju PAN.

Parysek, J. J., Wdowicka, M. (2002), 'Polish Socio-economic Transformation: Winners and Losers at the Local Level.' In: *European Urban and Regional Studies* 9: 1, pp. 73–80.

Riley, R. (2002) 'Embeddedness and the Tourism Industry in the Polish Southern Uplands: Social Processes as an Explanatory Framework.' In: *European Urban and Regional Studies* 7: 3, pp. 195–210.

Stawarska, R. (1999), 'EU enlargement from the Polish perspective.' *Journal of European Public Policy* 6(5), pp. 822–838.

Stenning, A. (2000), 'Placing (Post-)Socialism: The Making and Remaking of Nowa Huta, Poland.' In: *European Urban and Regional Studies* 7: 2, pp. 99–118.

Szymánska, D., Matczak, A. (2002), 'Urbanization in Poland: Tendencies and Transformation.' In: *European Urban and Regional Studies* 9: 1, pp. 39–46.

Weltrowska, J. (2002), 'Economic Change and Social Polarization in Poland.' In: *European Urban and Regional Studies* 9: 1, pp. 47–52.

Wever, E. (ed.) (2003), *Recent urban and regional developments in Poland and the Netherlands.* Papers presented at the Dutch-Polish economic geography seminar in Utrecht, 12–16 November 2001. Utrecht: Koninklijk Nederlands Aardrijkskundig Genootschap.

Czech Republic

Ministry of Economy (1996), *Housing in the Czech Republic. National Report for the United Nations Conference on Human Settlements – Habitat II.* Prague: Ministry of Economy of the Czech Republic – Housing Policy Department.

Musil, J. (1992), 'Recent changes in the housing system and policy in Czechoslovakia: An institutional approach.' In Turner, B., Hegedus, J., Tosics, I. (ed.), *The Reform of Housing in Eastern Europe and the Soviet Union*, pp. 62–70. London: Routledge.

Reiner, T.A., Strong, A.L. (1995), 'Formation of land and housing markets in the Czech Republic.' In: *Journal of the American Planning Association* 61: 2, pp. 200–209.

Strong, A.L., Reiner, T.A., Szyrmer, J. (1996), *Transformations in Land and Housing: Bulgaria, The Czech Republic and Poland.* New York: St. Martin's Press.

Sykora, L. (1996), 'The Czech Republic.' In: Balchin, P. (ed.), *Housing Policy in Europe*. London: Routledge, pp. 272–288.

Sykora, L., Simonickova, I. (1996), 'The development of the commercial property market and its impact on urban development in the context of transition from a command to a market economy: the case of the Czech Republic and Prague.' *EMERGO – Journal of Transforming Economies and Societies* 3: 3, pp. 74–88.

Slovakia

Buček, M. (1999), 'Regional Disparities in Transition in the Slovak Republic.' In: *European Urban and Regional Studies* 6: 4, pp. 360–364.

Slovenia

Cerne, A. (1997), 'Regional development in the context of spatial planning in Slovenia.' In: Einzelhandelsentwicklung, pp. 97–127.

Hungary

Baukó, T., Gurzó, I. (2001), 'Dilemmas in Agricultural and Rural Development in Hungary: The EU Accession Partnership and the SAPARD Programme.' In: *European Urban and Regional Studies* 8: 4, pp. 361–369.

Böröcz, J. (2000), 'The Fox and the Raven: The European Union and Hungary Negotiate the Margins of "Europe".' *Comparative Studies in Society and History* 42, pp. 847–875.

Hajdú, Z. (1994), 'European challenges and Hungarian responses in regional policy.' *Proceedings of the International Conference 'European Challenges and Hungarian Responses in Regional Policy'*, Pécs, 4–6 November 1993. Pécs: Centre for Regional Studies, Hungarian Academy of Sciences.

Horváth, G. (1999), 'Changing Hungarian Regional Policy and Accession to the European Union.' In: *European Urban and Regional Studies* 6: 2, pp. 166–177.

Keresztély, K. (2002), *The role of the state in the urban development of Budapest*. Pécs Centre for Regional Studies of Hungarian Academy of Sciences.

Kiss, E. (2004) 'Spatial Impacts of Post-socialist Industrial Transformation in the Major Hungarian Cities.' In: *European Urban and Regional Studies* 11: 1, pp. 81–87.

Kiss, P. (2001), 'Industrial Mass Production and Regional Differentiation in Hungary.' In: *European Urban and Regional Studies* 8: 4, pp. 321–328.

Locsmándi, G. (2001), *Urban planning and capital investment financing in Hungary.* Budapest: Open Society Institute.

Nagy, E. (2001), 'Winners and Losers in the Transformation of City Centre Retailing in East Central Europe.' In: *European Urban and Regional Studies* 8: 4, pp. 340–348.

Nagy, G. (2001), 'Knowledge-based Development: Opportunities for Medium-sized Cities in Hungary.' In: *European Urban and Regional Studies* 8: 4, pp. 329–339.

O'Relley, Z.E. (2001), 'From totalitarian central planning to a market economy: decentralization and privatization in Hungary.' In: *The Journal of Private Enterprise*, 17: 1, pp. 111–123.

Timár, J. (2001), 'Introduction: Uneven Development in Hungary.' In: *European Urban and Regional Studies* 8: 4, pp. 319–320.

Timár, J. (2004), 'What Convergence between what Geographies in Europe? A Hungarian Perspective.' In: *European Urban and Regional Studies* 11: 4, pp. 371–375.

Timár, J., Varadi, M. (2001), 'The Uneven Development of Suburbanization during Transition in Hungary.' In: *European Urban and Regional Studies* 8: 4, pp. 349–360.

Zetti, I. (2002), *La città post socialista: il caso di Budapest fra globalizzazione ed eredità passate.* Firenze: Alinea.

Cyprus

Godfrey, K.B (1996) 'The evolution of tourism planning in Cyprus: will recent changes prove sustainable?' In: *The Cyprus review*, 8: 1, pp. 111–133.

PART I
EAST-WEST PERSPECTIVES ON SPATIAL PLANNING AND URBAN DEVELOPMENT IN THE ENLARGED EU

Chapter 2

Spatial Development and Territorial Cohesion in Europe

Klaus R. Kunzmann

The Demise of the European Project?

In June 2005, referenda in France and the Netherlands brought the European Union into a crisis, a severe crisis. The subsequent quarrel over the European budget poured additional oil on the fire. The ambitious project of a strong Europe, successfully challenging the United States of America and Asia, particularly China, has to be overhauled. On the surface, the conflict is simply between two extremes: a Europe of nations in a free market economy, and a Europe which is governed by a huge bureaucracy in Brussels determined to control almost everything, from the size of cucumbers to European competition laws. In reality, the conflict encourages everybody to articulate vested concerns, justifying the critics with selective information about wrong-doings and mismanaged funds. There are those who criticise the European Agricultural Policy as an outdated model and urge the European leaders to dedicate more funds to research and development. Others defend the preservation of the European landscape and identity through agriculture, in order to maintain a longer term global competitiveness. They argue that by maintaining the various European regional identities and cultures, regional liveability and embeddedness, preconditions for attracting qualified labour, are guaranteed. In reality, the conflict is simply between those who trust the market more than state intervention and those who have confidence in an enlightened state when it comes to guide longer term spatial development.

Future spatial development in Europe, in Western and in Eastern Europe, will very much depend on the compromise line between the two positions, and on related territorial policies, whether they are European, national or regional. Obviously, experience shows that there is no alternative to the market economy. The question is rather how socially responsible such market policies should and can be in times of global competition, how much European or national policies, attempting to redistribute economic wealth within Europe find political and popular support. The challenges are obvious. Under conditions of a globalised economy, market focused policies will strengthen the centre of Europe against the periphery, the densely populated metropolitan city regions against rural and semi rural in-between-regions.

And these in-between-regions are those regions which are beyond the 100 km commuting distance to the metropolitan cores.

There is much evidence that social polarisation will further increase within both territorial categories. Hence there is a need for appropriate policies which aim to cushion the most negative social impacts. In city-regions as in peripheral and in-between-regions. To bring such development trends to a halt, state intervention is required in order to guarantee the access to basic public infrastructure. However, the degree to which the negative social and spatial implications of market forces can be tackled depends on the acceptability of equity goals and the willingness to share. The negative referenda in France and the Netherlands – and referenda in Germany or Italy or Sweden would not have brought any other outcome – demonstrated that the willingness to share is limited, that nation-centred policies are still dominating policies in the interest of Europe as a whole. The complexity of European policies in times of global economic competition is extremely difficult to handle and even more difficult to communicate in 21 languages to the people in the 25 countries of the European Union. Hence much effort still has to be undertaken to steer the ambitious European project in between the cliffs of regionalism, nationalism and century-long socio-cultural traditions on the one side, and non-spatial macro-economic policy options, competition laws and the global trade regulatory system on the other.

In 1957, in Rome, when Europe was established as an ambitious peace project, one could not have foreseen how far this project would have come in 2005. Nobody could expect that only 16 years after the fall of the wall in Berlin, most Central and Eastern European countries would already have joined the ambitious European project. Although there is a backlash in 2005, it is just a temporary one, and it will force all the intellectual leaders between Lisbon and Helsinki to explore more appropriate corridors in order to continue the process of European political integration and territorial cohesion, as the objective is called in the European Constitution.

Challenges for Spatial Development in Central and Eastern Europe

Under conditions of globalisation and technological change, the integration of regional and national economies into the competitive European market will cause considerable challenges. It will require long-term public action in the states of Central and Eastern Europe. National policy makers will have to mainly address six challenges: growing interregional social and spatial polarisation, structural change and industrial modernisation, agricultural divergence, insufficient transport infrastructure, environmental protection, conservation of the cultural heritage, and last but not least, the brain drain of the qualified labour force.

- Social and Spatial Polarisation: all over Central and Eastern Europe, social and spatial polarisation is rapidly growing. In the aftermaths of the demise of the socialist system in most central city regions spaces of affluence are visible. In gentrified inner cities, attractive suburban communities and in leisure regions the *nouveau-riche* upper class is showing its wealth. In contrast, those who did not succeed in benefiting from the new market economy are forced to remain in the run-down pre-fabricated social housing schemes at the urban fringe of agglomerations, where Western supermarkets and fast food chains absorb their small incomes. In the long run – and one generation may not be sufficient – the accession to the European Union will undoubtedly raise everybody's standard of living. However, during that long transition period the gap and the tensions between the winners and likely losers of the post-socialist open market system will grow. The political implications of such social and spatial polarisation are difficult to envisage and to handle.

- Structural Change and Industrial Modernisation: In some industrial branches (e.g. automotives) Central and Eastern Europe has become a favourite territory for West European corporations such as Volkswagen, Audi, Peugeot or Citroen. Attracted by cheap qualified labour, a strong work ethic and a growing market, West European corporations and firms have been heavily investing in strategic and easily accessible locations, where favourite regional traditions and qualifications guarantee speedy profits. Central and Eastern Europe, with the proximity to the large markets of Western Europe has much potential to compete with China and South East Asia, once these low wage regions will experience rising production costs as a consequence of social policies. Elsewhere, uncompetitive old industries are vanishing and their productions sites are falling idle. There, the transfer of experience of brownfield redevelopment policies in industrial regions of the Ruhr, Northern France or Great Britain would make sense, though it would have to be done without the generous financial support of The European Union.

- Agricultural Divergence: Agriculture in Central and Eastern Europe will experience growing divergence. While small subsistence farming will remain a feature in remote rural regions or as a weekend activity in urban agglomerations, large-scale industrial agriculture will gradually emerge, wherever soil conditions are favourable and agricultural land can be acquired and managed by capital intensive agro-investors. Certainly, it will take a generation until the demand for healthy food will bring about a new generation of farmers, who, as in Western Europe, are producing specialised farm products for middle class households and for export. Hence, for along time to come, the territories of Central and Eastern Europe will feature a heterogeneous mix of agricultural production modes, as the much disputed and very controversial traditional European Agricultural Policy will not alter the endogenous structures over night.

- Insufficient Transport Infrastructure: The interregional transport network in Eastern Europe is insufficient. The existing highway and train networks are inappropriate to accommodate all the new logistics flows between Northern and Southern, as well as Central, Eastern and Western Europe. High speed trains linking the larger agglomerations with the transportation hubs in Western

Europe do not yet exist. In some regions even pre-war rail connections were faster than today. And the airports require considerable efforts of modernisation to meet the requirements of the post-industrial European society. To bring the interregional transport networks in Central and Eastern Europe up to West European standards requires a huge multi-billion Euros programme. A few more decades and international financial resources will be needed to modernise the transport system in Central and Eastern Europe. Consequently, priorities will have to be set. Thereby it will not be easy to balance the market-driven demand for interregional European connections and the need to link rural and peripheral regions to the central hubs of the European networks. One can foresee that economic justifications will favour the densely populated urban centres over their rural hinterlands.

- Conservation of the Cultural Heritage: One of the precious assets of Central and Eastern Europe is the cultural heritage of their cities and landscapes. Untouched by forces of modernisation and profit-seeking investors over more than half a century, most cities and rural villages in Central and Eastern Europe have maintained much of their charm and character. This is an extremely valuable potential for both regional and international tourism and for citizens who can afford to leave the run-down housing estates at the urban fringe. However, such development will require a complex set of urban and spatial policies, ranging from home ownership programs and property tax reviews over quality control of tourist development to careful state intervention into property markets. The spectrum of necessary action at all tiers of planning and decision-making will have to range from awareness rising campaigns to financial support for investors in the urban heritage.

- Brain Drain of Qualified Labour Force: There is a growing mismatch between the number of highly skilled professionals in Eastern Europe and the number of better paid jobs available in the countries. Though to different degrees in single regions, there is much evidence that qualified labour force in the accession countries is being courted by institutions and industries in western Europe to leave and work temporally or permanently outside of their home regions. This is true for medical doctors for bankers as well as for engineers and computer scientists. They head to Great Britain or Germany, to Finland or Norway, where such qualifications are scarce or not available any more outside the larger urban agglomerations. Even if it is just a phenomenon of a longer transition period, the consequences for regional development are considerable. The brain drain will considerably affect peripheral and *in-between regions* in Eastern Europe. It will make it more and more difficult to sustain minimum public services in rural regions and small country towns. Distances to be covered for benefiting from such services will gradually increase and the rural exodus will continue with all its known backlash effects.

These challenges suggest that the paradigm of balanced spatial development, as promoted by the European Spatial Development Perspective (ESDP), could also be an appropriate *Leitbild* for future policy action in Central and Eastern Europe (CEC 2000). More than ninety per cent of the contents of the document, though written for the Europe of 15, are equally valid for the regions and states in Central and Eastern Europe.

The ESDP, a Paradigm for Spatial Planning in Europe

Spatial planning is an essential means to preserve European identity, in the West as in the East of Europe. It has a holistic dimension, it relies on local and regional endogenous cultures, and it is communicative by nature. Though spatial planning, as a rule, has limited political power and lacks efficient tools for implementation, it is a perfect arena for regional dialogue and learning. Despite all criticism and shortcomings, the ESDP, approved 1999 in Potsdam by the member states of the European Union (CEC 1999), has become an important policy document for spatial development in Europe. What is the importance of the document?

- First, the ESDP sets European-wide normative goals and principles of spatial planning at the regional and national levels. In particular, balanced spatial development and polycentricity are promoted as key concerns of spatial development. Based on experiences in Germany and the Netherlands, where polycentric development has been a dominant feature since spatial planning became a policy field after the second World War, the ESDP is aiming to transfer such experience to the whole of Europe. It is an effort to contain uncontrolled metropolisation and counteract short-sighted market driven concentration processes. Referring to the ongoing process of spatial concentration in Europe, however, one could easily argue that the ambitions as postulated in the ESDP contradict mainstream policies in the European Union. Indeed, most European sector policies, following mainstream market forces, favour, though not explicitly, the concentration of economic activities in a limited number of metropolitan regions.
- Second, the existence of the ESDP underlines the importance of the spatial dimension in sectoral planning. Given the weak position of spatial planning as a future-oriented policy field in most European countries, the document is at least a manifesto which stresses the role of space in sectoral policies such as transport, agriculture or energy. Only few institutions engaged in such sectoral policies will in the end use the document to check and eventually review the spatial implications of their activities and programmes.
- Third, the ESDP demonstrates the considerable communication power of the European Union. The document is one of the most circulated documents on spatial planning in Europe. It is available in all languages of the member states of the European Union, except those of the new members in Central and Eastern Europe. As a consequence, the ESDP has become a powerful pan-European source of information on principles of spatial development. It has

the quality of a textbook on how spatial planning at the regional and national levels should be done.

- Fourth, without the ESDP, the European Spatial Planning Observatory Network (ESPON) would never have been initiated. When working on the document it became soon apparent that reliable comparative space-related data in Europe are not available. While Eurostat could provide economic and social data the collection of spatial and urban data had been neglected. In order to obtain such data the member states agreed to establish and finance a network of national observatories with a small co-ordinating office in Luxemburg. This network has been asked to compile comparative information on spatial development trends in Europe and to commission transnational studies in areas where appropriate spatial information was not available (Davoudi 2005).
- Fifth, the ESDP contributed much to the justification of the various INTERREG programmes, which became a key instrument of the European Commission to promote interregional communication and exchange. Over the years, the INTERREG programmes have attracted much local and regional interest. The many interregional projects which have been supported by the Commission under the INTERREG label have contributed much to highlight the importance of the spatial dimension of regional development. They have brought together hundreds of regional planners, managers and policy makers across regions and nations who were able to learn from each other and promote a sustainable Europe as a common project (Schäfer 2003).
- Sixth, the ESDP, apart from being a significant employment initiative for internationally minded planners, has been very instrumental in bringing together European planners beyond their respective national academic and professional milieus and a few international networks. The document has triggered multiple debates in the international planning community on the nature and the rationale of spatial planning. More than once, such debates led to the formation of transnational research teams and networks who joined their forces for applications to European basic and applied research programs. In addition, the ESDP inspired journal and book editors to initiate special volumes on planning issues of European importance.
- Last, but not least, one could argue that the ESDP legitimises the role of the public sector in guiding spatial development with all its underlying social, cultural and environmental ambitions and concerns, which market forces tend to neglect. Although this has never been expressed explicitly, it is an essential dimension of territorial policies. Without a strong and efficient public administration, initiating continuous discourses on spatial development, influencing spatial policies by guiding private investment to appropriate locations, spatial cohesion in Europe cannot be achieved.

In the absence of the availability of hard instruments, spatial planning – compared to policy fields such as agriculture or transport – is by nature a rather weak policy area. In the post-industrial information society, however, spatial planning could be a valuable agenda-setting policy arena for sustainable development. The reasons are:

- The European space with all its cultural and scenic assets will be a key to the future economy of the continent. Spatially relevant policies, with all their knowledge of spatial conditions are crucial in defending amenities against uncontrolled economic growth and uncoordinated, non-sustainable infrastructure development.
- With its information and communication power and its strong concern for cities and regions, spatial planning reaches citizens all over Europe as it addresses problems of regional identity, cultural traditions and quality of life.
- The process of formulating a follow-up document to the ESDP is an important catalytic element to foster a European-wide discourse on a sustainable Europe. In no other policy arena, the multiple dimensions of sustainable development will be discussed in such a comprehensive way.
- The discourse on spatial planning at the European level will have some influence on European sectoral policies, such as agriculture, transport or competition policies. In the absence of any spatial framework and principles sector policies tend to neglect the likely spatial implications on cities and regions.
- Spatial planning promoted at the European level will encourage national governments to follow suit, either to elaborate and provide national concepts before the European Commission launches European-wide proposals, or to react to European proposals from a national position. Both requires strong national and regional spatial perspectives.
- Spatial planning efforts at the European level will require more up-to-date information on spatial development trends in Europe and on the requirements for spatial guidance and intervention. Consequently more transnational and comparative spatial research is necessary and will be justified.

Such reasons suggest that there should be a next ESDP or at least a document following up the ambitious goals of the Potsdam document (Kunzmann 2005).

Do We Need Another European Spatial Development Perspective? And What About Central and Eastern Europe?

At present, it seems, there is little enthusiasm and willingness at the European and national levels to invest much effort, time and money into European Spatial Planning. Apart from a clear commitment to continue the INTERREG Programme beyond 2006, no explicit efforts are being undertaken to update or extend it to the Central Eastern European territories. As a matter of fairness at least, the contents, aims and principles of the ESDP should be communicated to the planners and decision-makers of the new member states in their respective native language. However, there are good reasons not to sacrifice spatial planning to mainstream market-oriented economic growth policies.

At present, however, in mid-2005 there is no political commitment to produce precisely such a follow-up document, at least not one with such a denomination. What is in the pipeline is the preparation of a document to be launched by the EU Ministers for Spatial Development in 2007 when Germany will hold the Presidency of the European Union. It will be a document labelled '*Territorial State and Perspectives of the European Union*', building explicitly on the ESDP, though replacing 'space' by 'territory.' In this forthcoming document, '*territorial capital', 'territorial cohesio*n' and '*territorial development policies'* will be the politico-administrative buzz words. The ministers, together with the European Commission have already endorsed a scoping document and a summary of political messages for 'an assessment of the Territorial State and Perspectives of the European Union towards a stronger territorial cohesion in the light of the Lisbon and Gothenberg ambitions' (Conference of EU Ministers of Spatial Planning 2005). Three policy objectives for strengthening territorial cohesion will be elaborated in this document:

> Improving the strength and diversity/identity of urban centres/networks as motors for territorial development in Europe; improving accessibility and territorial integration in the Union, preserving and developing the quality and safety of Europe's natural and cultural values and developing sustainable urban rural-linkages. A special challenge in this respect is [...] to strengthen the territorial capital of areas with a weak economic structure or physical or geographical handicaps in an EU perspective, including their links to the potentially strong EU areas.

These are great ambitions and the 2007 document will be an impressive account of spatial planning rhetoric. How the ambitious goals articulated in the document can be implemented or at least infiltrated into partially contradicting sectoral policies will still have to be seen. One requirement will certainly be fulfilled: the forthcoming document will cover all 25 states of the European Union, hence include the new accession countries in Central and Eastern Europe.

One problem will remain: From what can be guessed from the scoping document the new territorial cohesion report will address the full complexity of territorial cohesion in Europe. This may lead to a document full of spatial/territorial development dreams and jargon, which, unless communicated to key decisions-makers and regional multipliers, may become sort of a paper tiger. Even if the focus is on 'territorial capital', it will certainly not argue that the development of the territorial capital of Paris or Munich will have to be slowed down in order to promote the territorial capital of peripheral or in-between-regions.

What could be a way out of the dilemma? While maintaining the European dimension, it might make sense to first agree on reducing the complexity of the future document to make it more down-to-earth and more readable for policy makers who are not familiar with the specialised jargon of spatial planners. The reduction of complexity could be done by focussing on selected spatial themes, themes which either reflect important European challenges or spatially relevant political concerns, such as knowledge industries, immigration or peripheral border regions, in-between-regions, medium sized cities, or regions with second home development pressure. There is one additional rationale for reducing the complexity of the new document by cutting its agenda into more easily digestible pieces: it will then be much easier to enrich the thematic documents by illustrative maps. In contrast to earlier versions of the ESDP, and as a consequence of some resistance by individual member states, the final document did not contain maps. A European territorial development perspective where space is reduced to symbolic pictograms is not really convincing. It does not use the power of images (Baudelle 2005, Duehr 2003). As the totally overrated though very popular 'Blue Banana' map has demonstrated, simple cartographic images are good means of triggering dialogues, particularly in multi-lingual policy environments. Words cannot substitute maps. Space or territory has a geographical dimension.

The next step is to address the most urgent challenges of European spatial development in Europe. This would require seeking a consensus regarding those challenges, even if the perspective differs from South to East and from West to North. One thematic area which may soon find political acceptance is the future development of those regions in Europe where accessibility is poor and basic infrastructure is eroding, and where, due to the absence of scenic beauty or cultural assets, even tourism and second homes development are not flourishing. Such regions are mainly situated in the periphery of Europe or in between the metropolitan European growth zones. Although no one openly dares to speak about it, such regions are 'looser regions' of globalisation, structural change and European spatial integration. However, to remain realistic, these regions can only be the target of compensation policies cushioning the negative impacts of regional decline by guaranteeing a minimum of basic public infrastructure. Whether Europe as a whole or national governments will have to maintain such basic infrastructure will be the outcome of political negotiations. A future territorial document dealing with this theme can only present a few success stories of spatial development, although all over Europe efforts are being made to identify effective ways to address the challenges of demographic and economic decline in such regions.

In the forthcoming document one could also focus on the spatial impacts of immigration on cities and border regions as one significant and highly sensitive policy field of urban development in Europe. Or one could address the spatial interrelationships and spatial implications of European transport development. Given their political sensitivity and controversies, such themes, however, may not be helpful in demonstrating the necessity and efficiency of European spatial development in media-dominated political environments. By contrast, one could explore less controversial spatial policy arenas which more likely validate spatial planning as a policy field. Water protection and flood control could be such themes, or tourism and coastal development. In these policy fields success stories are easier to find and easier to communicate than failures, unless one relies on the effects of threats and disasters, which the mass media tend to communicate with fervour.

In the end, the choice of themes will depend on the Council of Ministers of Spatial Planning and on the Commission in Brussels, who have to jointly decide on the future of territorial policies in Europe. They have to agree on which theme the future document should focus on within its ambitious and complex policy agenda. Given the power of the Structural Funds, the Commission can easily define thematic priorities. It remains important to promote spatial planning or/and territorial development as an instrument to sustain public discourse on the future of space in Europe, and to find political commitment beyond election campaigns.

Outlook: Soft Landing for Territories in Eastern Europe?

Whether the European Constitution will be accepted or not by all member states in the end, will not have much influence on the future day-to-day work of the Commission and the European Parliament. Although territorial cohesion is a key goal in the Constitution, the Commission is unlikely to expand its present European policy fields any further. While support to the new countries in the enlarged community will be favoured, present policies to speed-up and cushion structural change in the 'old' member states will gradually phase out. Most likely, subsidiarity will rather rule the game in the future. And the proposed focus on 'territorial capital' in the forthcoming 2007 document points to this direction. The 'No' to the constitution and a stagnating process of European expansion will offer new chances for deepening European integration. Spatial Planning in its old meaning or in its new outfit as territorial development is an ideal instrument to sustain continuous dialogue for sustainable European development in the face of globalisation. This, however, requires down-to-earth approaches in order to avoid that spatial planning or territorial cohesion remains a holy grail (Doucet 2005).

For the accession countries in Eastern Europe the explicit 'No' of some Western European states is a chance rather than a burden. Now they will have time to develop their own ways of spatial development based on their endogenous resources. Time is key to learning the lessons from failures and successes in Western Europe, and to adapt to the challenges of market and technology driven globalisation.

References and Further Reading

Baudelle, G. (2005), 'Figures d'Europe: Une Question d'Ìmage(s),' *Norois: Environnement, Aménagement, Société*, 1(194), pp. 27–48.

Bengs, C. and Böhme, K. (eds) (2004), *The Progress of European Spatial Planning*. Nordregio 1998, 1.

CEC (Commission of the European Communities) (1999), *European Spatial Development Perspective: Towards Balanced and Sustainable Development of the Territory of the EU*, Luxembourg.

CEC (1995), *Cohesion and the development challenge facing the lagging regions*. Regional Development Studies, No. 24, Luxembourg.

CEC (2000), *Spatial perspectives for the enlargement of the European Union*. Regional Development Studies, No. 36, Luxembourg.

Conference of EU Ministers of Spatial Planning (2005), *Scoping document and summary of political messages for an assessment of the Territorial State and Perspectives of the European Union towards a stronger European territorial Cohesion in the light of the Lisbon and Gothenburg ambitions.* 20/21 May 2005, Luxembourg.

Damette, F. (1997), 'Wie steht es um das Europäische Raumentwicklungskonzept?' *EUREG Europäische Zeitschrift für Regionalentwicklung*, 6, pp. 17–21.

David, C.-H. (2005), 'Zur Konvergenz der nationalen Raumordnungspolitiken Frankreichs und Deutschlands im Post-EUREK Prozess.' *Raumforschung und Raumordnung*, 63(1), pp. 11–20.

Davoudi, S. (2005), 'ESPON-Past, Present, and Future.' *Town and Country Planning* 74(3), Special Issue on European Spatial Planning, pp. 100–102.

Doucet, P. (2005) *Territorial Cohesion of Tomorrow: A Path to Co-operation or Competition.* Paper presented to the AESOP-Conference in Vienna, July 13–17, 2005.

Duehr, S. (2003), 'Illustrating Spatial Policies in Europe,' *European Planning Studies*, 11(8) pp. 929–948.

Duehr, S. and Nadin, V. (2005), 'The European Agenda and Spatial Planning in the UK.' *Town and Country Planning* 74(3), Special Issue on European Spatial Planning, pp. 83–85.

Faludi, A. (2000), 'The European Spatial Development Perspective – What next?' *European Planning Studies* 8(2), pp. 237–250.

Faludi, A. (2001), 'The Application of the European Spatial Development Perspective: Evidence from the North-West Metropolitan Area,' *European Planning Studies*, 9(5), pp. 663–675.

Faludi, A. (ed.) (2002a), *European Spatial Planning*, Lincoln Institute for Land Policy, Cambridge, MA.

Faludi, A. (2002b), 'Positioning European Spatial Planning.' *European Planning Studies*, 10(7), 897–909.

Faludi, A. and Waterhout, B. (2002), *The Making of the European Spatial Development Perspective: No Masterplan* (The RTPI Library Series), London: Routledge.

Faludi, A. (2004), 'The Open Method of Co-ordination and 'Post-regulatory' Territorial Policy.' *European Planning Studies* 12(7), pp. 1019–1033.

Faludi, A. (2005), 'Territorial cohesion; an unidentified political objective.' *Town Planning Review* 76(1), pp. 1–13.

Fauldi, A. (2005), 'Polycentric territorial cohesion policy.' *Town Planning Review* 76(1), pp. 107–118.

Jensen, O.B. and Richardson, T. (2004), *Making European Space: mobility, power and territorial identity.* London: Routledge.

Klein, R., Kunzmann, K.R. von Malchus, V., Tönnies, G. und Wolf, K. (1997) 'Comments on the Draft of a European Spatial Development Perspective (ESDP).' *EUREG Europäische Zeitschrift für Regionalentwicklung*, 6, pp. 37–41.

Krätke, S. (2001), 'Strengthening the Polycentric Urban System in Europe: Conclusions from the ESDP.' *European Planning Studies*, 9(1), pp. 105–116.

Kunzmann, K.R. (1984a), 'Eine Raumordnungskonzeption für Europa? Methodische und inhaltliche Annäherungen.' In: Akademie für Raumforschung und Landesplanung (Hg.), *Ansätze zu einer europäischen Raumordnung.* Forschungs- und Sitzungsberichte, Bd. 155, Hannover, 53–72.

Kunzmann, K.R. (1984b), 'Gedanken zur Erstellung eines Raumordnungskonzeptes für Europa.' *Berichte zur Raumforschung und Raumplanung* (Wien), 28. Jg., Heft 1, 3–10.

Kunzmann, K.R. (1984c), *The European Regional Planning Concept.* Council of European, Regional Planning Study Series, No. 45 (Appendix), Straßburg 1983, pp. 28–39.

Kunzmann, K.R. (1982), 'The European Regional Planning Concept.' *Ekistics*, 49(294), pp. 217–222.

Kunzmann, K.R. (1998), 'Spatial Development Perspectives for Europe 1972–97.' In: Bengs, C. and Böhme, K. (eds), *The Progress of European Spatial Planning.* Nordregio 1998, 1, pp. 49–59.

Kunzmann, K.R. (2005), 'Does Europe really need another ESDP? And if yes, how should such an ESDP + look like.' Manuscript of a lecture, given at the Dipartimento Archittetura e Pianificazione, Politecnico di Milano, 10 December 2003.

Schäfer, N. (2003), 'Ansätze einer Europäischen Raumentwicklung durch Förderpolitik – das Beispiel INTERREG.' Schriften zur Raumordnung und Landesplanung, Bd 14. Augsburg, Kaiserslautern.

Sykes, O. and Shaw, D (2005), 'Tracing the Influence of the ESDP on Planning in the UK Town and Country Planning.' 74(3), Special Issue on European Spatial Planning, pp. 108–110.

Williams, R.H. (1996), *European Spatial Policy and Planning.* London: Chapman.

Chapter 3

EU Enlargement and the Challenges for Spatial Planning Systems in the New Member States

Simin Davoudi

Introduction

In May 2004, ten new countries joined the European Union (EU). These are the Baltic states of Estonia, Latvia and Lithuania; the 'Visegrad'[1] countries of Poland, Hungary, the Czech Republic and Slovakia; the Mediterranean islands of Malta and Cyprus; and the only Balkan country, Slovenia. Romania and Bulgaria will follow in 2007 if they reform fast enough. The social, economic and spatial impact of the EU enlargement is considerable both for the new members and the Union as a whole. As in any other major transformation, the enlargement of the EU has led to strains and challenges as well as opportunities and rewards.

For the Union as a whole, a key spatial planning-related concern is the impact of enlargement on EU territorial cohesion. As illustrated by the *Third Report on Economic and Social Cohesion* (CEC 2004), in the context of an enlarged Union the challenge of achieving cohesion will be of a different magnitude, as the disparities in the enlarged EU are greater than ever. Whilst enlargement has increased the EU population by 20 per cent, it has added only 5 per cent to its Gross Domestic Product (GDP). Some 92 per cent of the population (i.e. around 73 million people) in the new Members States live in regions with a GDP per head of below 75 per cent of the EU 25 average, and over two-thirds in regions where this is under half the average. The speed with which the new states will catch up with the old ones depends on their annual growth rate. So, if a growth of 1.5 per cent a year above the EU 15 average is to be sustained, as has been more or less the case since the mid-1990s (CEC 2004: viii), it will take some 70 years for the new members to draw level with the rest of the Union (The *Economist* 2003: 4). The growth and convergence rates will of course vary from one country to another. For example, the Baltic countries have experienced a growth rate of 5 per cent or more in recent years and have hence the potential to converge more rapidly.

---

1   After the declaration they signed in 1991.

Hence, although the new member states have a recent shared history of being ruled by communist regimes and, partly because of that, suffered from similar structural problems, including obsolete and inefficient infrastructure, they will face their entry to the EU with a highly diverse range of cultural, social and even economic backgrounds. As regards the latter for example, the variations in GDP per head are wider between the new member states than between the countries of the EU15. Moreover, the disparities in these countries tend to be wider than across the existing Objective One regions. As a result of theses variations, the impact of enlargement on spatial planning systems in the new member states will undoubtedly be different, despite the harmonising effects of the EU regulations.

Therefore, any attempt to provide a one-size-fits-all account of such impacts runs the risk of over-generalisation and over-simplification. Nevertheless, it is the aim of this account to highlight three key challenges which will confront the spatial planning systems in almost all new member states, due not only to their entry to the EU but also their opening up to the free market and global economy some ten years or so before that. The first one is the rapid spatial restructuring of these countries at the national and regional scales. The second one is the rising tensions between the demand for development and growth, in particular infrastructure development, and the concern for the environmental protection. The third one is the trajectory of institutional restructuring and in particular the formation of new regional governance.

Spatial Restructuring and the Rise of Mono-Centric Development

One of the key challenges faced by the planning systems in the new member states is the emerging spatial restructuring which is characterised by a tendency towards mono-centrism and the widening of regional disparities. In the last ten years, most of these countries have witnessed a concentration of economic development and population in a small number of large metropolitan areas and particularly the capital cities. This has not only added to the concentration of congestion and environmental pressures in these areas, it has also widened the social and economic gaps between the regions.

Joining the EU is likely to accelerate the restructuring processes in these countries as they will become the main recipients of the EU structural funds in the near future. As was the case in most of the Cohesion Countries,[2] in the new member states only the major urban centres, and particularly the capital cities, have the critical mass, the infrastructure and the institutional capacity to absorb the EU resources and to deploy them effectively to achieve economic growth and higher living standards (Davoudi 2004). This differential capacity to mobilise EU resources to achieve further growth will lead to further uneven development. The experience of Ireland, following her entry to the EU, may well be repeated in the new member states. The impressive

---

2   Spain, Portugal, Greece and Ireland.

rate of economic development in Greater Dublin Area in the 1990s has gone hand in hand with a widening and deepening of regional disparities in the country (Davoudi and Wishardt 2005).

Within the new Member States the regional disparities, measured by GDP per head, is wider than in the Cohesion Countries and it has increased significantly since the mid-1990s, i.e. since the introduction of market economies (CEC 2004). In these countries, population growth and economic activities have already begun to gravitate towards capital cities such as Budapest, Prague, Riga, Tallinn and Ljubljana (Hall 2004). The only exception is Poland where there are a number of large metropolitan areas to rival Warsaw.

In the Czech Republic, for example, regional disparities have increased considerably after the state's policy of re-distribution came to a halt. Large metropolitan areas, such as Prague, Brno, Plzen, and Ceske Budejovice, have since witnessed faster growth as a result of their diversified economic base, highly skilled labour force, better cultural attractions and more efficient infrastructure. Prague in particular has pulled away from the rest of the country. Although its population represents 12 per cent of the total population, its average GDP generated between 1995 and 1997 reached 186 per cent of the country's average. In 1997, GDP per capita in Prague reached 122.5 per cent of the EU average (Illner 2001). Prague's rapid economic growth, however, has not been enjoyed elsewhere. Some of the worst hit metropolitan areas, such as Ostrava, Usti nad Labem regions, have suffered from declining economic activities in coal mining and heavy engineering sectors as well highly polluted environment.

This over-concentration of economic activities in Prague, and a few other large urban centres, has led to the intensification of the urbanisation process, where the entire metropolitan area, as opposed to an individual town or city, has become the focus of population concentration (Illner 2001). Smaller settlements in the vicinity of large cities, such as Prague, have witnessed major population growth in the last ten years. This, plus an underdeveloped housing and property market has led to extensive commuting over large distances, often on a weekly or monthly basis, whilst rural areas continue to face population loss.

Similar trends are taking place in Hungary where the Central Hungary Region and the agglomeration area of Budapest in particular is attracting population, foreign direct investment and economic growth. As in Prague, this has led to increasing suburbanisation. The GDP per capita in Budapest stood at 190 per cent of the national average in 1998. While the gap between Budapest and two other major counties, i.e. Gyor-Sopron-Moson and Fejer (in Western and Central Transdanubian Regions respectively) has reduced in the second half of the last decade, the disparities between the capital and the rest of the country, particularly places such as Nograd County (in the Northern Hungary) has continued to widen (Szigetvari 2001).

Slovakia, another Visegrad country, is also confronted with a restructuring process which favours large urban centres and particularly capital cities. Bratislava has been developing much faster than the other parts of the country in the last ten years, with its GDP gradually approaching the EU average. This has been accompanied with an intensification of suburbanisation due to the construction of residential districts in small suburban settlements around major cities. Kosice, capital of Eastern Slovakia, is following a similar trend to Bratislava (Pasiak et al. 2001). In Slovakia, the post-socialist restructuring has led to another major challenge for spatial development, that of the reduction in employment in agricultural industry and hence further loses of population in, and marginalisation of, rural areas.

Even Poland, which entered the post-socialist transformation with a well-balanced urban system, has since experienced a degree of regional polarisation. In 1997, the GDP in Warsaw region grew substantially to reach over 200 per cent of the national average while the GDP in the eastern voivodship of Zamosc dropped to mere 58 per cent of the national average. Whilst at the beginning of the 'transformation shock' (1989-1992) the more developed regions lost some of the strength of their economic position, they were able to take full advantage of the new, open economy afterwards. As a result, the big urban agglomerations have been able to maintain their leading position (Gorzelak 2001). Warsaw itself has become home to over 30 per cent of all foreign direct investment in Poland as well as the largest number of headquarters of international and national companies. Gorzelak (2001: 314) argues that metropolitan growth and restructuring in Poland has led to a new phenomenon: the metropolises are developing their links with other big cities in Europe and other major urban areas in the country at the expense of weakening of their ties with their own hinterlands. Whilst this may be true in terms of the ability of the hinterland to provide high quality economic input, it is not true in terms of provision of land for residential and recreational activities; i.e. accommodating the spatial growth of the metropolis. It is anticipated that by 2005, the regional disparities in Poland will be even greater than today and the GDP per capita in Warsaw municipal region will exceed the EU average.

All these examples clearly show that a key challenge facing strategic spatial planning in the new member states is to achieve a more balanced territorial development across the country without returning to the traditional regional policies of re-distribution. The challenge is to develop the indigenous potential of the underdeveloped regions without endangering the growth potential of the economically buoyant ones. According to the ESDP,[3] this challenge can be met by promoting polycentric development. Although elsewhere I have questioned the use of polycentric development as a panacea for solving regional problems (see Davoudi 2003), I strongly believe that the absence of a sound and coherent national spatial strategy can exacerbate such imbalances, as the Irish experience has shown. The rapid economic growth in Ireland took place in the context of infrastructure deficiencies and a lack of strategic spatial framework for the country as a whole. As the regional problems became more acute, the need for a spatial framework and effective planning policies became apparent. This led to the preparation of the National Spatial Strategy for Ireland whose central goal is to achieve a balanced regional development.

---

3   European Spatial Development Perspective published in 1999 in Potsdam.

The new members are fortunate as they have the opportunity to learn from the experience of the Cohesion Countries as well as from the widespread dissemination of the ESDP's principles, which have now been translated into some 14 languages including those of several new Members.

Lessons can also be learned from those Cohesion Countries that have experienced different spatial development trends. In Spain, for example, structural aid seems to have contracted the effects of polarisation due to a more balanced distribution of economic activity and settlements across the country. Among the new member states, Slovenia provides another example. Here, strategic planners have begun to formulate their national spatial strategy to guide their future spatial development in the course of an anticipated rapid growth assisted by the EU Structural and Cohesion Funds. The preparation of a national spatial strategy is a positive step towards providing a more effective framework for steering future developments and avoiding excessive concentration of activities in one or a few major urban centres.

Economic Growth and Environmental Protection

An overriding objective in all new member states is to achieve a high and sustained rate of economic growth, increased employment and better standards of living. However, this desire for economic growth has to be balanced against the need for environmental protection. Hence, a major challenge for strategic spatial planning for the new member states is to achieve such a balance by mediating between the contested interests over the use and development of land.

Economic growth as well as the integration into the EU requires good transportation and communication links between the key nodes. The new entrants to the EU have been building their economies largely on the infrastructure inherited from the socialist times. Not only they are inadequate and in need of replacement, they also often fail to connect the key nodes. For example, there is no fast line between the Baltic capitals yet and the one which is planned will take a decade to build. There is no direct train from Riga to Tallinn and the train to Vilnius takes a roundabout route. This is partly the legacy of the former Soviet Union's railway system that connected the cities of far-flung territory with Moscow first. However, the European Commission's plan to fund projects totalling some €220 billion to improve communications across the enlarged Union, including a new motorway from Gdansk to Vienna; another from Hungary across Romania reaching the Black sea at Constanta; and a third one across Bulgaria, linking Sofia with Thessaloniki, will change the picture dramatically (The *Economist* 2003: 6).

As regards the internal links, many Central European Countries, particularly Poland, are seriously short of decent roads. Poland has fewer kilometres of road than Slovenia which is one-tenth of its size (op cit), and spends only 19 per cent of its public funding on infrastructure, much less that Cyprus which spend 78 per cent of its public funding (CEC 2004: 172).

However, the challenge for strategic spatial planners is to ensure that the demand for new road building is not met at the expense of loss of environmental assets. It is also important that new developments are accompanied with investment in more sustainable modes of transport such as the revival of the declining rail networks. Rail carries 40 per cent of freight in Central Europe compared with 14 per cent in the EU 15 (The *Economist* 2003: 6). However, as regards passengers, the rapid suburbanisation and the rising levels of commuting coupled with inadequate public transport systems has already led to bottlenecks and congestion around many large urban areas and particularly capital cities. It is alarming that private car ownership rose by 80 per cent across central Europe in the 1990s and more than doubled in the Baltic countries. Congestion is only one consequence of the increasing use of private car; the other is the environmental pollution. There is also a third consequence; that of increasing road accidents. Latvia leads Europe for road-deaths. Enlargement has in fact increased the EU-wide average rate for road deaths per person by 5 per cent (op. cit.).

In addition to new roads, there is a need to bring the water, sewage and waste systems up to the EU standards and close those nuclear power stations in Lithuania and Bulgaria which do not comply with EU regulations. For the next 20 and 30 years, the new member states will be the site of major new infrastructure development. It is of paramount significance that such developments take place in the context of a strategic spatial framework that aims to achieve a more balanced and sustainable pattern of growth at the national and regional levels. Spatial planning has a major role to play not only in promoting territorial cohesion but also in pursuing policy integration between various policy sectors.

Emerging Regional Governance

A third related challenge for spatial planning is the nature and quality of institutional context within which the planning systems operate. The preparation for EU entry has given the new member states the motivation and the role models for establishing not only the market economies but also for reviving democratic institutions. One of the first indications of decentralisation was devolution to local governments. Whilst this brought decision-making processes closer to the communities, some argue that it initially led to fragmentations. In Hungary, for example, the radical reduction in the roles and functions of counties, which represent a thousand years old tradition of governmental system (Szigetvari 2004), left a policy vacuum at the intermediate level of governance, leading to a lack of coordination between local governments.

In Poland, too, the post-1989 political transformation led to the elevation of the role and power of local governments as well as the abolition of counties. However, in both cases, counties have now been effectively replaced with elected regional governments. It is important to note that the creation of these new regions has been influenced by the prospect of joining the Union. The boundaries of the newly formed regions have been drawn in such a way that matches the requirements of the

Commission both in terms of statistical convenience and eligibility for Structural Funds. Hence, the outcome of the territorial sub-division of the new member states has been the creation of NUTS II regions, which are bigger than the traditional counties in Hungary, Poland and Czech Republic, for example.

Despite the technocratic rationale behind such administrative sub-divisions, some commentators argue that the new regions provide 'a better unit for regional programming and strategic planning, since the sphere of influence in several activities crosses the county border' (Szigetvari 2004: 300). Indeed, it is at this scale where strategic spatial planning can be formulated most effectively. In countries such as Poland and Hungary the new framework of regional policy supported by directly elected regional governments has marked the beginning of the integration of sectoral and regional policies. However, the degree to which spatial planning is integrated into regional policy is questionable. As Illner (2004: 281) argues, 'regional policy in the Czech Republic is conceived predominantly as a set of economic measures promoting economic development, rather than a comprehensive instrument including also spatial planning. The latter represents a separate system with its own rules and procedures'.

It is important to note that the creation of regional level has not taken place in all new member states; for example, in Slovakia "there is a missing dimension of strategic and long-term planning and a strategic vision of community development" (Pasiak et al. 2004: 337). This is partly manifested in the absence of a regional self-government and the strengthening of the role of state at the expense of the devolution of functions to territorial and local government. Addressing regional disparities will prove very difficult in the absence of effective regional governance.

Another significant post-1989 institutional change which has been further enhanced by the prospect of joining the EU is the emergence of new forms of governance, consisting of not only governmental bodies but also other public and private stakeholders, and the formation of numerous associations and cross-border co-operations. The latter has been substantially facilitated by various EU funding streams. The implications for the planning systems in the new member states are significant because, such associations have the potential to provide the much needed democratic arenas for strategic coordination of various policy sectors, and for seeking consensus over critical yet contentious decisions about the management of spatial change. This will be particularly the case in the next decade where these countries will be subject to pressures of rapid and colossal growth.

Concluding Remarks

When the Berlin Wall fell, people in the former Eastern European countries began to celebrate, among other things, the end of planning as they knew it, i.e. central top-down, command and control planning imposed by communist regimes. Hence, for some time, it was quite unpopular to discuss the need for and the benefits of strategic spatial planning systems. In many new member states, the post-1989 rapid economic growth and its associated massive expansion of the built environment, particularly in the major urban centres and capital cities, took place within limited strategic frameworks for spatial planning and land use regulation. The entry to the EU will further accelerate this growth and amplify its social and environmental and spatial consequences. It is therefore crucial that the future transformations of urban and rural areas are managed in such a way that is environmentally more sustainable, socially more equitable and spatially more balanced. The existence of an effective spatial planning system at both strategic and local levels is central to achieving this goal.

References

CEC (European Communities) (2004), *A new partnership for cohesion, Third Report on Economic and Social Cohesion*, Luxembourg: Office for Official Publications of the European Communities.

Davoudi, S. and Wishardt, M. (2005), 'The Polycentric Turn in the Irish Spatial Strategy.' *Built Environment*, 31: 2, pp. 122–133.

Davoudi, S. (2003), 'Polycentricity in European Spatial Planning: From an Analytical Tool to a Normative Agenda.' *European Planning Studies*, 11: 8, pp. 979–999.

Davoudi, S. (2004), 'Territorial cohesion: An agenda that is gaining momentum.' *Town and Country Planning*, 73: 7/8, pp. 224–227.

Gorzelak, G. (2001), 'The regional dimension of Polish transformation: seven years later.' In: Gorzelak, G., Ehrkich, E., Faltan, L. and Illner, M. (eds), *Central Europe in Transition: Towards EU Membership*. Warsaw: Scholar Publishing House.

Hall, P. (2004), 'A Taste of Battles to Come.' *Town and Country Planning*, 73: 1, pp. 8–9.

Illner, M. (2001), 'Regional development in the Czech Republic: The 1993 scenario revisited.' In: Gorzelak, G., Ehrkich, E., Faltan, L. and Illner, M. (eds), *Central Europe in Transition: Towards EU Membership*. Warsaw: Scholar Publishing House.

Pasiak, J., Gajdos, P. and Faltan, L. (2001), 'Regional patterns in Slovak development.' In: Gorzelak, G., Ehrkich, E., Faltan, L. and Illner, M. (eds), *Central Europe in Transition: Towards EU Membership*. Warsaw: Scholar Publishing House.

Szigetvari, T. (2001), 'Regional development in Hungary.' In: Gorzelak, G., Ehrkich, E., Faltan, L. and Illner, M. (eds), *Central Europe in Transition: Towards EU Membership.* Warsaw: Scholar Publishing House.

The *Economist* (2003), 'When East Meets West: A survey of EU enlargement.' 22 November.

Chapter 4

The European Union and the European Cities: Three Phases of the European Urban Policy

Susanne Frank

Introduction

What does the term 'European urban policy' mean? Is there any such thing at all? This question is highly debated, and whenever asked, different people give very different, even diametrically opposing answers. Depending on the institutional affiliation and interests, the range of answers reaches from 'No, there is no European urban policy, but there definitely should be one' (probably a representative of *Eurocities*) to 'Yes, there is, but there better not be' (certainly a civil servant of the German government).

The discrepancies in the assessments and ratings are not exclusively, but particularly due to the fact that there is no supranational responsibility for urban policy within the European Union (EU). Thus, the European Commission is lacking a legal and legitimising foundation for 'official' actions in this field. Nevertheless, for many years it has in fact intensely been striving towards gaining influence on the development of European cities – even though without having yet succeeded in obtaining more formal competencies which had been its more or less openly claimed objective during the 1990s.

The difficult and painful processes of social restructuring during the 1980s triggered a growing interest in urban policy on the part of the European institutions. In the beginning, the European Parliament was the driving force. Soon after the first direct election in 1979, Members of Parliament started to point out the crisis-prone economic, ecological, and social situation of many cities and particularly of the conurbations. Several times, the European Commission was summoned to be proactive in these political fields (Paulus 2000: 97), and when Jacques Delors entered upon office in 1985 it began to manifest a growing interest in urban issues. The typical procedure was to consult a wide range of experts and lobbyists. Authorized academics but also many lobbyists and even local authorities encouraged the Commission to systematically attend to urban development and to develop an independent European urban policy (Burton et al. 1986, Cheshire et al. 1987, 1989, Eurocities Conference 1989, Parkinson et al. 1992, Hachmann 2000, Petzold 2000).

But right from the start, the process was accompanied by criticism. It was mainly the member states and regions who interpreted the intervention of the EU in urban affairs as an infringement of the subsidiarity principle.

In this paper, I neither intend to go over the individual stages nor the different player constellations in the struggle for an independent urban dimension of European politics.[1] It is rather to be shown that the supranational efforts to define a European urban policy were of a relatively short but alternating history. This history can be divided into three phases and associated with three political fields: During the 1980s urban-related activities aimed primarily at the environmental policy. In the 1990s urban issues became part of the cohesion policy, formally remaining there until today. But as to its content the city as a topic becomes more and more part of the economic and competition policies. In addition to that, I would like to argue that these shifts express a change in the understanding of the 'European city' which is after all the target of the interventions. And finally, the altered understanding of the city itself reflects in itself a change of the idea of the European Union as a political community altogether.

These theses ground on the analyses of central documents accounted for by the Commission, on the observation of four conventions on (the future of) the European urban development as well as on several interviews and numerous informal – but hence highly informative – conversations and discussions with players within the field of European urban policy.

The Invocation of the 'European City'

The Commission's first approach to urban development was performed in the field of the environmental policy. The Single European Act of 1986 declared the preservation and protection of the environment for the first time as one of the Community's fields of action. The Commission used this gain in competence as a gateway to urban activities: It postulated that Community action for the protection of the environment 'must include the potential of such action within urban areas' (COM 1990: 1). In these terms, the responsible Directorate-General for the Environment presented the *Green Paper on Urban Environment* in 1990 which aimed at the 'development of a Community strategy for the urban environment' and submitted hereunto a multitude of measures and initiatives. The Green Paper is the central document which outlines and represents the supranational urban activities of the 1980s.[2]

It is a remarkable document for several reasons. First, by publishing the Green Paper the Commission expresses for the first time openly and officially its interest in putting the urban topic on the European agenda. Second, it is striking that the

---

1   See the minute reconstruction in Paulus (2000), also see Tofarides (2003).

2   Green papers are discussion papers published by the Commission on a specific policy area. Primarily they are documents addressed to interested parties – organizations and individuals – who are invited to participate in a process of consultation and debate. See http: //europa.eu.int/documents/comm/index_en.htm (17.02.2005).

Commission still deals quite unbiased with the difficult and highly explosive issue of legitimising its intervention, that is its lack of a political mandate at the local level. Thus, it says succinctly in the foreword: 'Of course, urban issues remain primarily in the responsibility of local and national authorities; however, a Community action which identifies common problems and enhances the exchange of know-how in the course of searching for optimal solutions is feasible while allowing for the subsidiarity principle at the same time' (KOM 1990: 5). A supranational responsibility concerning urban development is being claimed quite bluntly in the introduction.[3] The Commission declares what is not to be denied: The majority of the Community's (sectoral) policies and initiatives (for example transport, energy, social affairs) exert, directly or indirectly, influence on urban areas (COM 1990: 1, 39). Therefore, the various sectoral policies at the Community level would have to 'take due account of the problems of the urban areas and converge into a Community strategy for Europe's cities' (ibid.: 2): An 'evaluation of urban-friendliness' at the supranational level ought to ensure that adverse effects on the urban environment, particularly through EU policies, are to be avoided (ibid.: 39).

Thirdly, the broad definition of the term 'environment' as used in the Green Paper will strike at least the German reader. The Green Paper deals with the term far beyond its narrow meaning of harmful substances, noise, waste, water quality and so on. Urban environment is interpreted in a very comprehensive way – economic, political, social, and cultural issues are included as a matter of course. Thus, it is rather a 'green urban paper' than an 'urban Green Paper' (Kunzmann 1990: 846). The Commission uses the term 'environment' as a Trojan Horse, first, to position itself as a new player in the field of urban policy and, second, to introduce its vision of a genuine European urban development to the political debate. In fact, the Green Paper claims no less than to propose a 'specifically European approach to cities' (COM 1990: 8).

In order to do so, it is taking stock of the situation and problems of the European cities at the end of the 20th century. In large parts, the specific qualities of the European city are stressed and virtually praised. On the one hand, the meaning of the dynamics, creativity, and innovation of urban centres for an economic, social, cultural, and political development of Europe is emphasized. The city, as the cradle of European civilization and democracy, is supposed to be the 'place where direct participation is possible and increasingly practiced, and where the individual can

---

3   It is striking that the level of member states or the national level respectively is hardly to be found at all in the document. Klaus R. Kunzmann reasons that it seems 'as if there were only two levels of action in Europe: the Community and the cities', which does not exactly reflect 'the true political proportions' (1990: 849). The claim for supranational responsibility in regard to urban environment is almost openly formulated. A clear-sighted commentator as Kunzmann foresees the Commission having to face questions like 'Which Directorate-General of the Commission ought to be authorized with a European urban planning policy? How to efficiently divide the tasks among the two relevant Directorates-General DG XVI [Regio, SF] and DG XI [Environment, SF], or are there good reasons *to implement an independent Directorate-General*?' (Kunzmann 1990: 846, accentuation SF).

develop most freely his sense of personal and civic value. It is not by accident that citizen, citoyen, cittadino, or Bürger denote the political sovereign in our languages' (ibid.: 9). Much space is given to the appreciation of the historical heritage and its meaning for a European identity: 'The historical character of our European towns and cities – their buildings, monuments, their squares, and street patterns – establishes an identity and sense of place specific to individual cities. Our cities are an important symbol of the Community's rich cultural diversity and its shared historical heritage. Interest in protecting a city's historical character is therefore not restricted to that city's own citizens' (ibid.: 46). That last sentence already reveals a right to competence which is again more blatantly claimed when the Commission calls upon the 'recognition of a European dimension of the historic and cultural heritage of our towns and cities' which ought to be the basis of Community action 'with reference to the urban environment' (ibid.: 35).

An immense endangerment of the European city is diagnosed along its outstanding social meaning. It is noted that urban exodus and urban sprawl, suburbanization, mass consumption and so on would destroy its specific characteristics and achievements. 'At stake is the quality of "civilization" in its most practical manifestations of economic, scientific, and social performance' (ibid.: 6). According to the Commission, the causes for 'urban degradation (...) are to be found in the way we organize work, production, distribution, and consumption, and in often rigid and outdated notions of planning' (ibid.: 26). Among many other developments, the functional zoning of the city (Charter of Athens) is criticized, further the destruction of local traditions by an architecture and urban planning oriented towards international ideals ('uniform banality of the international style'; ibid.: 30), the decline of the inner-cities due to the drift of trade and consumption towards the outskirts, the decay of public spaces, and social segregation (ibid.: 26ff). Contrasting this, it is called for a planning which 'takes as its model the old, traditional life of the European city, stressing density, multiple use, social and cultural diversity. Different social, professional and age groups living together also create the basis for a civil coexistence which is undermined by growing mutual ignorance and distrust' (ibid.: 29).

The patient is ill but not dead yet. For its recovery the Commission considers two unalterable elements as the 'crux of a European concept': First, 'the European city can still be saved' (ibid.: 8). Of course, there would be problems but they seem to be resolvable. It requires economic growth, though, to obtain the necessary financial means. In the eyes of the Commission the second non-deceivable basis is 'Europe's traditional commitment to what is now called "social cohesion"' (ibid.). European cities would have to provide 'equal and decent living conditions' and supply their inhabitants with work and perspectives (ibid.).

I am not discussing the Green Paper in order to critically analyse the positions and claims of the Commission regarding urban development – although there would be a multitude of reasons for this.[4] I would rather like to point out the comprehensive and

---

4   Among these are the superficial analysis of the 'urban degradation' which is traced back to incorrect planning – it is confounding that economic and political causes do not get

diverse ways in which the Commission refers to the topic of urban development in Europe. It does not at all reduce the city to its economic meaning. The Commission's intervention is based on a highly normative vision of an urbanized Europe, the 'Europe of cities'. Relevant keywords are *civil society, democracy, integration, participation, history,* and *identity*. The economic, the social, and the spatial cohesion are praised and glorified as the European city's specific qualities – it is quite easy to detect herein the 1980s apparition of an Americanisation menacing the European urban development.[5]

Struggling for the 'Social City'

The publication of the Green Paper was the prelude to an intensive process of consultations which came to an end at a conference in Madrid at the end of April 1991. A number of important urban and environmental actions at the European level resulted from it.[6] But by the time the Madrid results were published under the title *City and Environment* (COM 1994)[7] there had already been a shift of emphasis:

---

mentioned in this context at all; the naive – and partially tending towards a constructional-spatial determinism – faith in planning; also, the undiminished focus on economic growth; the harsh mixture of descriptions, opinions, valuations, and postulations (Kunzmann 1990: 847); the dangerously uncritical idealization and idyllization of the 'traditional life of the European city' and its overestimation as model for the future city; the provincial opposition to international architectural styles and so on.

5   A fourth particularity ought to be pointed out at the margins: Especially when compared to other EU-publications, the Green Paper's careful and elaborate design is striking. The cover-picture has not been chosen casually: It is adorned by a painting of dark demons circling banefully above the roofs of a city (it is part of the tenth painting of Giotto's world-famous fresco cycle at the Basilica San Francesco in Assisi). Also, two pictures inside the volume are fraught with allusions: The first chapter ('The Future of the Urban Environment') and the third chapter ('The Root Causes of Urban Degradation') are each introduced by threatening scenes of the movies *Metropolis* by Fritz Lang and *Brazil* by Terry Gilliams. Both movies are known as classics of dark and disconcerting visions of large cities. This systematic selection of pictures stresses the Commission's concern about the crisis-prone situation and the future of the European city.

6   Among them the implementation of the *EC Expert Group on the Urban Environment* in 1991, the *European Sustainable Cities Project* in 1993, and the *European Sustainable Cities & Towns Campaign* in 1994. The topic 'environment' holds a central position at the European level until today; see for example the communication *Towards a Thematic Strategy on the Urban Environment*. Especially in this document it is striking that the broad understanding of environment has decidedly narrowed: Today it focuses mainly on a sustainable urban development on ecological terms and particularly on urban 'health problems related to environment' (KOM 2004b).

7   The documentation of the convention's contributions (COM 1994b) witnesses and emphasizes one more time the comprehensive claim to and the meaning of the 'vision' of the European city to different players (see esp. 'Chapter 2: A European vision of the city', pp. 33ff).

The environmental policy had been superseded by the cohesion policy as the principal field of action in which an independent European urban policy was to be established.

That shift may be explained by two causes, regarding concept and strategy. First, meaning and effectiveness of the cohesion policy increased considerably after the reform of the structural funds (1988). The latter are used by the EU in order to achieve economic and social cohesion: They serve to promote measures reducing the imbalances between regions and/or social groups.[8] The structural policy's gain in competence becomes most apparent in the continually increasing financial supply in the 1990s (up to 35 per cent of the entire EU budget). The structural or cohesion policy is one of the very few supranational redistributive fields of politics. The fact that the Commission does have money to distribute considerably widens its scope of action and increases its influence on regional and local politics, especially in times of economic hardship: The allocation of money can be – and actually is – subjected to conditions regarding concept and procedure. Second, the cohesion policy is institutionally and symbolically elevated when the promotion of the economic and social cohesion is being embodied as one of the three principle objectives of the Union in the Maastricht Treaty – alongside the economic and monetary union and the single market (par. 2 and par. 158).

Further, the cohesion objective's explicit recognition as one of the Community's principle objectives is also of relevant meaning to the Commission because urgent calls for action can be asserted in its name. Thus, the cohesion objective serves the Commission to legitimate its engagement in urban (development) policies – which is not stipulated in the treaty. For this, the Commission develops an explanatory figure which is repeated innumerable times during the 1990s: 80 per cent of the European population live in cities. Thus the EU is the world's most urbanized region. Since cities are, on the one hand, centres of economic growth and innovation, one will have to support them in order to promote the European economy and competition. On the other hand, diverse problems like unemployment, poverty, exclusion, crime, traffic congestion, and ecological damage concentrate in the cities and especially in specific neighbourhoods. Thereby, other values and norms of the Union, namely integration and cohesion, are violated. As far as all nations and cities of the Union are affected by these problems a European dimension is diagnosed – regardless of local, national, and regional differences. Consequently, supranational measures are not only justified but imperatively recommended. Therefore, the Commission is

---

8   The *European Regional Development Fund* (ERDF) and the *European Social Fund* (ESF) are of primary importance for urban policy. The principle objective of the ERDF is to promote economic and social cohesion within the European Union through the reduction of imbalances between regions and/or social groups. The ESF is the principal financial instrument allowing the Union to realize the strategic objectives of its employment policy. Further, there are the *European Agricultural Guidance and Guarantee Fund* (EAGGF – Guidance Section) and the *Financial Instrument for Fisheries Guidance* (FIFG) (http://europa.eu.int/comm/regional_policy/funds/prord/sf_en.htm; 18.02.2005)

able to interpret its engagement in cities and neighbourhoods as contribution to the economic, social, and territorial cohesion policies.

By applying the cohesion objective to the level of cities and neighbourhoods during the second phase, the Commission increasingly focuses on smaller areas (Eltges 2002: 256). From now on, the supranational engagement is to be legitimated not only by the differences in the development and wealth of large areas, but also by those on a small and smallest spatial scale; next to the regional balancing there ought to be that of cities and neighbourhoods. 'That is to say: A second pillar is to be added to the cohesion policy' (ibid.).[9]

In numerous documents the Commission admits that, among other reasons, its very own economic and competition policies are responsible for the increase in social inequalities. Therefore, the EU policies themselves produce winners and losers within the social transformation process. The social gap keeps growing even in rich and prospering cities. In a publication on the Poverty 3 program – with the significant title *Social Europe. Towards a Europe of more solidarity: Combating social exclusion* – the estimate of the number of persons affected by poverty and marginalization at the beginning of the 1990s amounts to 50 million (KOM 1993: 3). According to the Commission, the increasing socio-spatial polarization and segregation would have caused a 'fear' of 'social exclusion becoming a characteristic of the European social landscape in the long run' (ibid.). Therefore, the Community would have to 'intensify its efforts of avoiding and fighting social exclusion in order to obtain the Europe of solidarity that the citizenship of the Union lays claim to' (ibid.). The Commission intends to primarily focus its efforts on the city and its neighbourhoods: 'The problems of the urban areas require greater and more broadly dispersed endeavours during the 1990s' (KOM 1991: 133).

The first cohesion report is presented by the Commission in 1996 and substantiates the Union's idea of a 'European solidarity' (COM 1996). The principle objective of the cohesion policy is seen in contributing to the maintenance and promotion of a 'European social model' which is based upon the values of the social market economy. 'This seeks to combine a system of economic organisation, based on market forces and freedom of opportunity and enterprise with a commitment to the values of internal solidarity and mutual support which ensures open access for all members of society to services of general benefit and protection' (ibid.: 13). As the report shows, the cohesion policy of the 1990s grounds on the conviction that the instruments of the market economy are insufficient to combat the increasing territorial disparities and social problems, that markets ought to be regulated, that market failure should get corrected, and that resources ought to be redistributed. In this respect, the cohesion policy implies a specific idea of Europe, namely a 'European model of society which is founded on the notion of the social market economy' (ibid.: 14) – and thus, it is directly competing with the neoliberal concept of politics and society: 'The 1988

---

9   By the end of the 1990s this claim is turned into applicable law for the 2000-2006 structural funds' period. Under the heading *Objective 2*, 'urban areas in difficulty' are for the first time recognized as an independent category for promotion.

cohesion policy reform has been a bedrock of the anti-neoliberal program. Though the immediate objective was to reduce territorial inequalities, its larger goal was to strengthen European regulated capitalism' (Hooghe and Marks 2001: 106).

Indeed, the Commission becomes very active in combating specific urban problems throughout the 1990s – all actions are carried out in the name of the cohesion policy and in order to preserve the 'European social model'. Already at the beginning of the 1990s the Commission starts experimenting with incentive measures for a *social development of cities and neighbourhoods*. In this context, the Community Initiative URBAN, committed to revitalize neighbourhoods in difficulty, is of particular meaning. It is divided into two periods: During the first term from 1994 to 1999 (URBAN I), 116 neighbourhoods with a total of 3.2 million residents were promoted; during the second and still ongoing phase from 2000 to 2006 (URBAN II) the program is applied to 70 cities and/or neighbourhoods. These programs helped to direct the focus of parts of the EU structural policy purposefully and explicitly on urban areas.[10]

URBAN is the ambitious (and at the same time ambivalent)[11] attempt to counteract the 'increasing tensions within European societies' which become primarily apparent 'in a high degree of social exclusion in a growing number of inner-cities and on the outskirts' (KOM 1994b: 6). The objective is to sustainably improve the living conditions in selected cities and neighbourhoods – by creating jobs, promoting human resources, improving the socio-cultural infrastructure as well as the environmental quality, and by the integration and networking of local players. The new quality of the social problems in the cities and neighbourhoods ought to be addressed by a new policy which, as known, can be described as area-related, integrated, innovative, promoting partnership, participatory, sustainable, and as gender-equitable. All efforts ought to focus on the needs, interests, and potentials of the affected neighbourhoods and social groups. Further, the program URBAN includes the support of city-networks in order to enhance the exchange of practical experience and know-how and to develop sustainable good and best practice models (ibid.: 7).

By implementing the Community initiative URBAN (and all other structural funds programs) the Commission pursues two objectives. On the one hand, it aims at a socio-integrative strengthening of 'areas where development is lagging behind', as required by the cohesion policy. On the other hand and at the same time,

---

10   Community initiatives are specific financial instruments of the supranational level which decides to a large extent on their use. They enable the Commission 'to use the means of the structural funds in a more flexible manner and to apply them rather to European than to national needs'. Within the scope of the cohesion policy the Community initiatives serve to promote innovative and experimental measures which are of special interest to the EU and of 'additional use' to the Community. This includes especially the development of transnational co-operations and exchanges as well as the set-up of networks among public and private players (Staeck 1997: 102).

11   See Walther (2002) and Walther and Güntner (2002) on ambitions and ambivalences of socio-integrative urban development policies.

URBAN ought to carry a new political instrument of control into the cities and neighbourhoods. With the implementation of URBAN the normative and political issue of 'good' governing at the local level and 'good' management of cities or neighbourhoods becomes relevant. To the EU, the structural funds are important instruments to provide the cities and neighbourhoods with incentives to adapt to the EU's ideas of *good governance* and also to experiment with new forms of governance. In order to receive financial support the Member Nations and/or sub-national units have to develop and implement so called operational programs which meet the maxims, guidelines, and modalities issued by the European Union – that is programs which correspond to the EU's ideas of good governance and of good city or neighbourhood management.[12] Thereby, the EU interferes to a large extent with national and local policies, structures, institutions, and traditions. Before rewarding with money, the EU demands to implement its norms – which often leads to a radical change in politics. In this respect, approaches like URBAN do not only express an altered understanding of 'urban renewal policies', but they fundamentally aim at a 'renewal of urban policies' at the same time (Franke et al. 2001). This process intends that local actions obtain a European dimension and vice versa, that European actions obtain a local dimension by focusing on cities and neighbourhoods.

URBAN fits into an overall concept of urban policies which the Commission explains in the 1997 communication *Towards an urban agenda in the European Union* (COM 1997). The document intends to formulate the challenges which European cities are facing, to develop strategies and guidelines for a joint way of dealing with them, and to start a comprehensive debate on urban issues. As far as the communication precisely states the (inner contradictions of the) guiding principles which structure the Commission's activities throughout the 1990s, it can be considered the peak of the second phase of the European urban policies and at the same time also as their final or turning point.[13]

The document starts like a thunderbolt: Beginning with the idea of European cities being the driving forces of economic and social development and at the same time having to deal with the social consequences of the rapid transformation, the Commission criticizes that many national policies' lack a vision to combat the problems: 'It is clear that new efforts are necessary to strengthen or restore the role of Europe's cities as places of social and cultural integration, as sources of economic prosperity and sustainable development, and as the bases of democracy' (COM 1997: 3). This model of an urban-based European democracy is praised as 'European

---

12   Also see the *White Paper on European Governance* (COM 2001).

13   Only a year later the Commission presents the communication *Sustainable Urban Development in the European Union: A Framework for Action* (COM 1998). It is the result of the wide and lively debates on the communication *Towards an urban agenda* and is, until today, the basis of discussions among the EU and its Member States, regions and cities. Though the *Framework* simply repeats many of the rationales, the variations and supplements indicate already a transition towards a new, third phase of European urban policies.

model of society' whose keywords are *equality of rights, equal opportunities, shared affluence, social security, solidarity,* and *integration.*[14]

At the same time the urban-based 'European model of society' is regarded as highly endangered: The 'erosion of the role of the city is perhaps the greatest threat to the European model of development and society and one which needs the widest debate' (ibid.: 8). Thereby, the Commission has defined its role within this process. It understands itself as the guarantor of this model of society and thus as the trustee of a historical heritage as well as of a future vision which it considers otherwise insufficiently protected and maintained. The Commission highlights that 'cities in the European Union are facing a number of common problems' which call for the acknowledgement of 'opportunities at the European scale to share and facilitate potential solutions' (ibid.: 3). Aware of numerous Member States opposing to a European competency regarding urban policies, it states clearly that 'this would not require additional powers at the European level. Rather, much can be achieved through a more focused approach using existing instruments at national and Community level' (ibid.). Albeit this, it lists a multitude of arguments for a strengthening of the European level in regard to urban issues. Thus, as a consequence of the statement that all EU activities have a direct or indirect bearing on the 'development and quality of live' in European cities, the Commission holds it necessary to occupy a(n) (institutionalised) 'complementary role' in addressing urban issues (ibid.: 14). It would be essential to coordinate and integrate the existing (national and common) policies and instruments more efficiently. In order to achieve this, the Commission intends to formulate standardized guidelines, to constitute general frameworks, to develop strategies, and to give incentives for development – all in terms of strengthening the 'European model of society'.

As many other documents of the 1990s, also this one is deeply penetrated by the knowledge of the tense relationship between the ideas and goals of the economic and the cohesion policy. Repeatedly it is conceded that not all of the Community's policies have a positive effect on the cities. 'The twin challenge facing European urban policy is therefore one of maintaining its cities at the forefront of an increasingly globalized and competitive economy while addressing the cumulative legacy of urban deprivation. These two aspects of urban policy are complementary' (COM 1997: 13). Thus, urban policy is facing two tasks: maintaining the cities at the forefront of an increasingly globalized and competitive economy and at the same time managing a difficult heritage resulting from the decline of the cities. Shortly thereafter, it is one more time stated perfectly clear: 'Economic progress which

---

14 For example, the 'crucial role (of cities) in underpinning a European model of society, based on equal opportunities regardless of gender and ethnic origin' is emphasized (ibid.: 3). Further down, the Commission advises the Member States to guarantee a common and area-wide access to and accommodation with services of public interest. These are said to be 'part of shared values in Europe' and 'are at the heart of the European model of society, since they further fundamental objectives of the European societies such as solidarity and equal treatment within an open and dynamic market' (ibid.: 15).

undermines the cohesiveness of urban areas is unlikely to be sustainable over the longer-term' (ibid.).

Promoting the 'Competitive City'

In the early 1990s, the political scientist Gilbert Ziebura considered the nation state's position at the beginning of a 'new global era which still has to define its organizing principles' (1993: 35f) as dilemmatic: 'On the one hand, it has to protect its society from considerable negative effects of the "globalization" process, but at the same time it has to do everything to open it for this process, to "limber" it up, and to guarantee its survival in an increasingly tough competition, because withdrawal or closure would tantamount to decline. Obviously it is swamped with this "balancing act"' (ibid.: 41).

This analysis of position is easily being transferred to the supranational level. The inner tension as outlined above characterizes particularly the cohesion policy and the European urban policy. As I have tried to point out, its second phase is characterized by an oscillation between the two conflicting goals of the Community: to set up a Common Single Market and enhance the competitiveness of the European cities on the one hand, and to promote economic and social cohesion within and amongst the cities on the other hand. The cohesion policy was supported by the adherents of the social market economy and against the ideals of neoliberal European politics. It gave fresh impetus to those who interpreted the structural as well as the urban policy as striving for solidarity and for a balancing of privileged and disadvantaged regions and/or social groups – within the framework of an urban-based European model of society which is founded on the notion of the social market economy. Programs like the Community initiative URBAN focus clearly on the needs and interests of the weaker ones and engage in achieving an economic, social, and territorial balance.

Below, I would like to point out that the emphasis has radically been shifting since the late 1990s. The understanding of the cohesion policy and, therefore, the role and meaning of the European urban policy have undergone a fundamental change.

As early as 2001, Hooghe and Marks spoke of a 'threat' (2001: 105) to the cohesion policy in its form of that time. Since the late 1990s the eligibility criteria and the enforcement regulations of the structural funds were gradually softened which the authors interpret as a shift in the balance of power between the adherents of a 'neoliberal' capitalism and those of a 'regulated' capitalism in favour of the former: 'For neoliberals, a Europe-wide social policy, tax policy or cohesion policy is undesirable because it constrains market competition' (ibid.: 109). The 1990s cohesion policy, committed to the idea of the social market economy, contradicts the neoliberal model of economics and society in several ways: firstly, because markets are purposefully intervened by the structural funds, and secondly, because there is redistribution among affluent and poor countries or regions. In addition to that, vertical and horizontal partnerships consign political and economic decision making to market-distant players. The consultation and participation of those who are potentially concerned by these decisions hinders, slows down, or prevents decisions considered as economically effective. Also, the economic criteria for efficiency and rationality would often be counter-effected by the partnerships' and networks' complex ways of functioning which always aim at consensus and balance of interests: '(Partnership) is grounded on principles of inclusiveness and consensus, rather than market competition' (ibid.).

What was years ago timidly described as a tendency by Hooghe and Marks, becomes entirely obvious during the current renegotiations of the structural funds for the cohesion policy's term 2007–2013: The structural dilemma described above is being solved in favour of economic fitness. The adoption of the so called Lisbon strategy elucidates this strategic shift. In March 2000 the European Council resolves upon an agenda on issues of employment in order to, as is often cited, make the Union 'the most competitive and dynamic knowledge-driven economy in the world'. On the EU web site it is elaborated: 'For the purpose of this strategy, a strong economy supports the creation of jobs and promotes social and ecological measures which on their part guarantee sustainable development and social cohesion'.[15] From now on, economic growth and increased competitiveness are the principle objectives all other political fields have to focus on and commit to.

Formerly a bulwark for those who do not favour the neoliberal model of economics and politics, the cohesion policy loses its function as opponent or merely as a corrective of the competition policy. Its alterations result in a fundamental change of the European urban policy. Although the urban (development) policy remains formally within the DG Regio and therefore still is part of the structural and cohesion policy, its motives and goals have undergone such a fundamental change that I would like to speak of another shift of focus: Within the EU, the topic city becomes increasingly an issue of the competition policy.

The political and strategic shifts become very obvious when comparing the first and the third cohesion report in which the Commission introduces its concept of an extended Union for the planning term 2007–2013. While the first cohesion report deals in detail with the fundaments of the European model of society and its inherent concept of the social market economy, the third cohesion report barely considers any on the fundamental values of the EU and of the structural policy. The first cohesion report unequivocally states, addressing the neoliberals: 'The Union's political goals of solidarity, mutual support and cohesion may be pursued through largely economic means, but (...) these goals, nevertheless, remain the irreducible ambitions which structure European society and help to determine its sense of identity' (COM 1996: 14), whereas the third report subsumes the cohesion objective almost bluntly under the idea of enhancing competitiveness. Thus, already the introduction (usually the place of grant and appealing words) defines the role of the cohesion policy as having to increase the economic growth of the Union. One passage expresses the henceforward subordinated meaning of the idea to promote and maintain the 'European model of society' as well as the new and

---

15   http://europa.eu.int/comm/lisbon_strategy/index_de.html, 03.10.2004.

altered tone of the document: 'Strengthening regional competitiveness throughout the Union and helping people fulfil their capabilities will boost the growth potential of the EU economy as a whole to the common benefit of all. And, by securing a more balanced spread of economic activity across the Union, it will reduce the risk of bottlenecks as growth occurs and lessen the likelihood of inflationary pressure bringing growth to a premature end. It will equally make it easier to sustain the European model of society and to cope with the growing number of people above retirement age and so maintain social cohesion' (COM 2004: viii). This and numerous other passages show that, in the eyes of the Commission, there are no longer any 'irreducible efforts', values, and objectives that would tantamount to economic efficiency and competitiveness.

What does this strategic readjustment of the structural and cohesion policy mean for the endeavours towards a European urban policy? My thesis is: In the context of the Lisbon strategy, the cities will be paid even greater attention to than before. Nevertheless, direction and focus of this interest will change fundamentally.

The hitherto existing Community initiative URBAN will probably not be continued. The Commission does not intend, though, to stop considering urban issues in particular or to stop committing to an approach of urban renewal. On the contrary: It intends to 'reinforce the place of urban issues by fully integrating actions in this field into the programs' (COM 2004: xxxii). The URBAN approach ought to be transferred into mainstream as URBAN+. But key promotion modalities and eligibility criteria of the Community initiative URBAN will probably be given up or become diluted.

Thus, the Commission's draft for the 2007–2013 structural funds decree designates each Member State to propose (by the beginning of the next planning period) a list of cities and urban areas which ought to be promoted and therefore be addressed by specific measures according to the programs. But in the future, it will be the Member States and their regions who decide upon the way they integrate the URBAN approach into their regional programs. Consequently, vis-à-vis the Commission, they are regaining in influence on the urban (development) policy. Of course, the control modalities connected to URBAN ought to be preserved, but one of the most important 'orthodoxies' of the past years will be softened: the postulated focus on the most disadvantaged or crisis-afflicted areas of a city. These changes within the concept of promotion become tersely obvious in a new and currently often cited wording which shifts the focus from the 'areas in need' to the 'areas of opportunities'. Apart from Objective-1, it is demanded to consequently direct the programs and initiatives towards the new priorities of the Community, namely towards the Lisbon Agenda.[16] Accordingly, structural funds interventions no longer ought to focus primarily on the want of the most indigent neighbourhoods and persons, but ought to be implied where specific possibilities

---

16 Formally, also the orientation towards the Gothenburg Agenda is demanded, but actually most documents and discussions refer to the Lisbon strategy only (see for example the Third Report on Economic and Social Cohesion, COM 2004a).

and potentials – in terms of economic growth and competency – are detected. It ought to be about the stimulation of 'prospects' or 'resources' respectively. 'What we want is to enhance competitiveness' – this sentence was often heard when best practice projects were introduced at the *Urban Development* section of this year's *Regio Open Days* in Brussels. And the head of the Urban Unit prophesied at the same meeting: 'Competitiveness will be the key word'.

The debate on the 'Guadeloupe problem' at the *Regio Open Days* reveals quite clearly the extent to which the promotion policies have already changed in regard to idea, concept, and ambience. A representative of the city of Newcastle criticized the taking into account of regions at the far outskirts or those featuring unfavourable geographical conditions – as it is stipulated in the cohesion policy even after its reform. 'What does Guadeloupe contribute to the Lisbon Agenda? What does the Lisbon Agenda contribute to Guadeloupe?' (Richardson 2004). Instead of focusing on an equalization of living conditions the promotion policy ought to concentrate on 'strong' cities and regions. Their boom would finally reflect on the 'less developed' ones. Many observers familiar with the European urban policy agreed upon the notion that this kind of questioning central values and principles of the cohesion policy would have caused protest or at least statements of disapproval a few years ago – today it can count on approval, even applause.

These debates outline the contours of a 'new European model of society' (Voelzkow 2000) in which cities and regions play an outstanding role. The governments' actions at national and supranational level no longer target at a well balanced internal welfare or at social equalization but at international competitiveness. The keywords are well known: from the *welfare state* to the *competition state*, from *welfare* to *workfare*. Social participation (also and especially) of disadvantaged neighbourhoods and groups of people ought to be achieved by their integration into the markets. 'It is being attempted to meet the objectives of social equalization and social integration (...) by promoting economy and employment' (ibid.: 512). Thus, also the understanding of solidarity changes: 'A protectionist and redistributing solidarity is (...) step by step being replaced by a solidarity oriented towards competition and productivity' (ibid.: 510, see Streeck 1999).

Among others, also many supporters of the European urban policy are aware of the fundamental changes of its concept, definitions, and objectives. Senior staff of the Urban Unit of the DG Regio realize, not without regret, that today's meaning of solidarity and inclusion differs completely from that of ten years ago when international lobby organizations like *Quartiers en Crise* or CECODHAS exerted great influence on the structure of the program. The head of the Urban Unit summarizes the change: 'Today, helping those with the most severe problems doesn't necessarily mean that help has to be directed only to the special spot, but can be understood much broader' (Regio Open Days, Sept. 2004).

Conclusion: The European Union and the 'European City'

European urban policy is of a short but agitated history. It was, is, and will be a combated, heterogeneous, and unstable field of politics. As shown above, the efforts to outline and establish an independent European urban policy change emphasis several times: They move from the environmental policy via the cohesion policy to the competition policy. Each of these shifts is accompanied by an altered understanding of the social meaning of the European city for the European society. This process can be described as a gradual narrowing of perspectives and a loss of the complex idea or 'vision' of the European city.

During the first phase, European politics focused on an idealized European city as the core of a European identity which needed to be (re-)stabilized. During the second phase, they dealt with the European city as the place where the 'European model of society', based on the values of the social market economy, was to be put in concrete form. During the third and still ongoing phase, such a comprehensive vision of the European city whose values exceed merely economic terms can no longer be perceived without further ado. But that does not mean that the urban dimension is becoming less important to the structural policy. Cities and regions, considered as sources of growth and innovation, are rather gaining in importance. Nor is to be stated that urban policies no longer integrate a 'European model of society' – on the contrary: European urban and regional policies play an important role in achieving a European model of society which is increasingly interspersed with neoliberal ideas. This one is obviously only interested in cultural, social, moral, normative aspects and problems of urban development if they can be capitalized for the purpose of the Lisbon Strategy.

In the course of this development, knowledge from the social sciences and political experiences, which played an important role in outlining European politics during the 1990s, is increasingly ignored: for example the often verified fact that economic growth is no cure-all for poverty and exclusion, but on today's conditions, goes along with and even produces and reproduces them. The same applies to the political insight that the very own economic, growth, and competition policies (likewise and especially) result in an increase in social and spatial disparities – that is why the combat against them has also always been understood as a question of social moral and solidarity. The shifts described will be of consequence for those 'disadvantaged' or 'problematic' cities, neighbourhoods, and social groups that 'flive more of' the European economy than they contribute to it.

The role of the urban dimension within the future cohesion policy is currently negotiated. But even members of the Urban Unit fear that mostly the big competitive cities will benefit from the new eligibility criteria as outlined in the Lisbon Agenda, and that other cities or weaker social interests will be further marginalized. The Urban Unit seems to stand up for maintaining the ideas of a socio-integrative urban development as part of the incentive measures. But if the Urban Unit in its function as coordinator of urban issues will continue to exist at all is not yet decided upon.

Translation: Birgit Helms

References

Burton, P., Forrest, R. and Stewart, M. (1986), *Lebensbedingungen im städtischen Europa*. Luxembourg: European Commission.

Cheshire, P. and Hay, D. (1989), *Urban Problems in Western Europe. An Economic Analysis*. London: Harper Collins.

Cheshire, P., Hay, D., Carbonaro, G. and Bevan, N. (1987), *Urban Problems and Regional Policy in the European Community*. Luxembourg: European Commission.

COM 1990/European Commission (1990), *Green Paper on Urban Environment* (http: //europa.eu.int/comm/environment/urban/pdf/com90218final_en.pdf, 24.02.2005).

COM 1993/European Commission (1993), *Social Europe. Towards a Europe of more solidarity: Combating social exclusion*. Luxembourg: European Community.

COM 1994/European Commission (1994), *City and Environment*. Luxembourg: European Community.

COM 1996/European Commission (1996), *Report on economic and social cohesion. First cohesion report* (http: //europa.eu.int/comm/regional_policy/sources/ docoffic/official/reports/repco_en.htm, 24.02.2005).

COM 1997/European Commission (1997), *Towards an urban agenda in the European Union* (http: //europa.eu.int/comm/urban/documents/d004_en.pdf, 24.02.2005).

COM 1998/European Commission (1998), *Sustainable Urban Development in the European Union: A Framework for Action*. Luxembourg: European Community.

COM 2001/European Commission (2001), *White Paper on European Governance*. Luxembourg: European Community.

COM 2004/European Commission (2004), *A new partnership for cohesion convergence competitiveness cooperation. Third report on economic and social cohesion* (http: //europa.eu.int/comm/regional_policy/sources/docoffic/official/ reports/pdf/cohesion3/cohesion3_conclusion_en.pdf, 11.02.2004).

Eltges, M. (2002), 'Die soziale Stadt und die Europäisierung der Stadtpolitik.' In: Walther, U.-J. (ed.), *Soziale Stadt – Zwischenbilanzen. Ein Programm auf dem Weg zur sozialen Stadt?* Opladen: Leske+Budrich.

Eurocities Conference (1989), *Eurociudades=Eurociutats=Eurocités=Eurocities: Documents and Subjects of the Eurocities Conference in Barcelona*. Brussels: European Community.

Franke, T., Löhr, R.-P. and Sander, R. (2002), 'Soziale Stadt – Stadterneuerungspolitik als Stadtpolitikerneuerung.' *Archiv für Kommunalwissenschaften*, 39 (2002), pp. 243–268.

Hachmann, C.J. (2000), 'EU-Programm zur Förderung einer dauerhaften Stadtentwicklung. URBAN II erfordert rasches Handeln.' *ZdW Bay*, 4 (2000), pp. 188–191.

Hooghe, L. and Marks, G. (2001), *Multi-Level Governance and European Integration*. Lanham: Rowman & Littlefield Publishers.

KOM 1990/Europäische Kommission (1990), *Grünbuch über die städtische Umwelt*.

Luxembourg: European Community.

KOM 1991/Europäische Kommission (1991), *Europa 2000. Perspektiven der künftigen Raumordnung der Gemeinschaft.* Luxembourg: European Community.

KOM 1993/Europäische Kommission (1993), *Soziales Europa. Für ein solidarischeres Europa: der Kampf gegen die soziale Ausgrenzung.* Luxembourg: European Community.

KOM 1994a/Europäische Kommission (1994a), *City and Environment.* Luxembourg: European Community.

KOM 1994b/Europäische Kommission (1994b), 'Mitteilung an die Mitgliedstaaten zur Festlegung von Leitlinien für die von ihnen zu erstellenden Operationellen Programme im Rahmen einer Gemeinschaftsinitiative für städtische Gebiete (URBAN).' *Amtsblatt der Europäischen Gemeinschaften vom 1. Juli 1994: 6/7.*

KOM 2004a/Europäische Kommission (2004a), *Eine neue Partnerschaft für die Kohäsion. Konvergenz – Wettbewerbsfähigkeit – Kooperation. Dritter Bericht über den wirtschaftlichen und sozialen Zusammenhalt.* Luxembourg: European Community.

KOM 2004b/Europäische Kommission (2004b), *Entwicklung einer thematischen Strategie für die städtische Umwelt.* Luxembourg: European Community.

Kunzmann, K.R. (1990), 'Durch Wiese und Wald in die europäische Stadtpolitik. Zum Grünbuch der Kommission der Europäischen Gemeinschaft über städtische Umwelt.' *Der städtetag*, 12 (1990), pp. 846–851.

Parkinson, M. et al. (1992), *Urbanisation and the Function of Cities in the European Community.* Brussels/Luxembourg: European Community.

Paulus, S. (2000), *'URBAN'. A Critical Case Study of the Formulation and Operationalisation of a Community Initiative*, PhD thesis. London: London School of Economics.

Petzold, W. (2000), 'Die Städtepolitik der EU und Möglichkeiten der Strukturfondsförderung 2000–2006.' In: Führungsakademie der Wohnungs- und Immobilienwirtschaft e.V. (FWI) (ed.), *Städtebau als integriertes Gesamtkonzept für die Zukunft*, Hamburg: FWI.

Richardson, K. (2004), *The role of major cities in regional EU programmes* (unpublished lecture at the Regio Open Days in Brussels), September 2004.

Staeck, N. (1997), *Politikprozesse in der Europäischen Union.* Baden-Baden: Nomos.

Streeck, W. (1999), *Competitive Solidarity: Rethinking the 'European Social Model'.* Köln: MPIfG Working Paper 99/8.

Tofarides, M. (2003), *Urban Policy in the European Union. A Multi-Level Gatekeeper System.* Aldershot: Ashgate.

Voelzkow, H. (2000), 'Regieren im Europa der Regionen.' *Informationen zur Raumentwicklung*, 9/10 (2000), pp. 507–516.

Walther, U.-J. and Güntner, S. (2002), 'Transnationales Wissen in Revitalisierungsstrategien – Informiert oder deformiert die Stadtpolitik der EU?', in Walther, U.-J. (ed.), *Soziale Stadt – Zwischenbilanzen. Ein Programm auf dem Weg zur sozialen Stadt?* Opladen: Leske+Budrich, pp. 265–276.

Walther, U.-J. (2002), 'Ambitionen und Ambivalenzen eines Programms. Die Soziale Stadt zwischen neuen Herausforderungen und alten Lösungen.' In: Walther, U.-J. (ed.), *Soziale Stadt – Zwischenbilanzen. Ein Programm auf dem Weg zur sozialen Stadt?* Opladen: Leske+Budrich, pp. 23–44.

Ziebura, G. (1993), 'Nationalstaat, Nationalismus, supranationale Integration. Der Fall Frankreich.' In: H.A. Winkler and H. Kaelble (ed.), *Nationalismus – Nationalitäten – Supranationalität.* Stuttgart: Klett-Cotta.

PART II
SPATIAL PLANNING AND
SUSTAINABLE DEVELOPMENT
IN THE NEW EU
MEMBER STATES

Chapter 5

Sustainable Spatial Development in Slovenia: Between Global Trends and Local Urban Problems

Kaliopa Dimitrovska Andrews

Introduction

Slovenia is a small country, both in size (20,273 km²) and population (2 million inhabitants), located at a principal crossroads in Central Europe. The richness of its cultural and historical experience is an important asset in Slovenia's potential for future economic and social development, and the diverse climate and variety of landscape (e.g. from Adriatic and Alpine to Panonian) provide ideal preconditions for a wide range of different economic activities and lifestyles.

A diverse industrial history, a tradition of openness to the world, and well-managed State economic policy all contribute to the fact that Slovenia is one of the most successful countries in transition from a socialist to a market economy. Per capita GDP is 9,760 USD, with a growth rate of 4.6 per cent. Slightly more than one half (56 per cent) of the total active population is employed in the service sector, 40 per cent in manufacturing and industry, and 4 per cent in agriculture. Territorial organization comprises only two levels: the national, and the local government authority. The process of regionalization is underway.

The population of the territory of Slovenia has grown by four per cent in the last two decades. The birth-rate is one of the lowest, both in Europe and the world. The positive growth of population in the past (until the mid 1980s) was to a large extent the result of the high migration of a young labour force from other parts of the former Yugoslavia. The age structure of the population and the size of households have followed the same patterns as elsewhere in Europe: the population is ageing, the size of households has become smaller (from 3.5 in 1971 to 2.8 in 2002), and the share of single-person households has increased (17 per cent vs 22 per cent).

Slovenia has a polycentric settlement structure based on eight regional development centres: only two of these cities have a population of over 100,000 (Ljubljana 265,881 and Maribor with 110,000), five have populations of between 20,000 and 50,000 (Kranj, Celje, Novo Mesto; Koper, Nova Gorica, Murska Sobota); 12 centres of higher importance with around 10,000 inhabitants, and 62 local centres. Generally, the settlement patterns of urbanization reflect the endeavours to stimulate small- and medium-size towns while reducing regional inequalities. More than half the population live in urban areas.

Slovenia has a total housing stock of 777,772 units (Census 2002), of which 52 per cent are located in urban settlements. With slightly over 110,000 dwellings, the capital city Ljubljana accounts for 28 per cent of the total urban housing stock. Approximately 37 per cent of the total number of dwellings are in multi-family, high-density residential blocks constructed mainly during the 1970s and 1980s (Dimitrovska Andrews and Sendi 2001). The share of homeownership is among the highest in Europe (92 per cent) (Dimitrovska Andrews and Černič Mali 2004). The majority of the housing in private ownership is owner occupied.

The expansion of the Slovenian cities and major towns in the period since the Second Word War has included all the basic types of urban growth and change (Dimitrovska Andrews 2000). The 1960s, 1970s, and the first half of the 1980s were characterised by the concentration of population and working places in towns and by decline in agrarian activity and depopulation in the rural areas. The period after 1985, and more extensively after 1991, was characterised by urban growth by substitution in the existing urban areas, involving demolition and reconstruction (brownfield development). There have also been processes of urban growth by 'satellite' extension, characterised by the development of new suburban areas within the functional urban regions of large cities: principally with dormitory neighbourhoods during the period 1970 to 1985, followed by mixed uses after 1985.

Impact of the European Union on Physical Planning

The predominant influence of the EU on spatial planning in the member states has occurred directly through several avenues (EC 1997):

- Legislation, especially harmonisation of the environmental laws,
- Policy documentation, on matters with a spatial dimension, especially concerning the Trans European Networks (TEN) infrastructure,
- Policy formulation and implementation, notably cohesion policy supported by the Structural and Cohesion Funds.

The EU Funding of Programmes has had a direct spatial impact on former socialist countries in the context of regionalisation, preparation of regional development plans and the establishment of regional development agencies for the organisation and review of structural fund spending.

As well as the relatively direct impact of the European Union through law, policy and funding, the European dimension is reflected in other ways, in changes to the planning systems in these countries, and indirectly their physical planning. Firstly, recent changes to planning systems show, to some extent, an increasing concern with strategic planning, not only at the regional level, which in part reflects the perceived growing importance of European integration, but also at the city level (e.g.

Prague, Warsaw, Ljubljana). Secondly, there is evidence of the impact of EU and other international policies through:

- the adoption of objectives, guiding principles and criteria for sustainable development in most of the city planning documents and policies (e.g. European Spatial Development Prospective (ESDP), the Green paper on the Urban Environment, Habitat Agenda, Agenda 22);
- the promotion of new planning methods and the exchange of know-how on city planning and management through networking of research institutes, city planning departments, city authorities and other important actors in the planning process;
- the operation of international real estate investment (foreign firms, loan activities of the European Investment Bank, World Bank and other international banks) that are dominant in city transformation and restructuring processes especially in Prague, Budapest and Warsaw – the most important gateway cities (Keivani et al. 2001).

National Urban Policies in Slovenia: Towards Sustainability

Spatial Planning System and Policies

Following EU guidelines, the reform of spatial planning and land use management in Slovenia has been addressed by the adoption of the new Spatial Planning Act (2003) and the Construction and Facilities Act (2003), as well as by adoption of numerous regulations on the basis of the EU Directives. The new planning system is expected to enable better flexibility of spatial documents, greater public participation within the planning process, and to establish the foundations of a spatial planning information system.

State policies in housing, employment and economy, education and training, health, social care and welfare are based on sector-specific Acts, mainly in the form of national programmes (e.g. National Housing Programme, Programme for Combating Poverty and Social Exclusion, National Programme for Environmental Protection). During the preparation phase of a national programme, there is usually an extensive public debate in which experts, NGOs, representatives of different institutions and organizations, as well as the lay public, can take part. National programmes are then adopted by Parliament according to a standard procedure. The implementation of policy programmes and measures is mostly the responsibility of State bodies, while the local government authorities and other actors (NGOs, private parties) are also involved to varying degrees. When responsibility for services is shared between the State and local authorities, programmes are very often developed in close cooperation between the responsible State and local government authority institutions. A local authority has to participate by co-financing programmes (adult education, public work programmes, local development programmes for example).

State   In spatial planning and management, the State prepares legislation, policies, and other instruments, prior to adoption by the National Assembly or the Government of the Republic of Slovenia. They define the spatial planning system and provide strategic spatial development objectives and guidelines. In addition to the provision of a framework for spatial planning at regional and local levels, the State also implements measures concerning spatial development activities and construction that are of national significance (e.g. utilities, power stations, motorway construction). The spatial planning documents at the national level are the Spatial Development Strategy of Slovenia, the Spatial Order of Slovenia, and the Detailed Plan of National Priorities. The State has the authority to monitor the legitimacy of spatial planning at lower levels and to take alternative action, if a local community fails to perform its tasks with respect to spatial planning and management. The State also has the responsibility to conduct and implement land policy, maintain the spatial data system, develop and encourage professional work in spatial planning, and participate in the matters of spatial planning and management at the international level.

Region   Slovenia has not yet formally established the regional administrative level. The only spatial planning instrument at the regional level is the Regional Spatial Development Plan, which is being prepared jointly by the State and local government authorities according to the principle of partnership. If this planning document is prepared in sufficient detail, it can replace the Spatial Development Strategies for the local government authorities involved. Regional Development Agencies have been established for the 12 statistical regions (Dimitrovska-Andrews and Ploštajner 2000) to prepare the regional programme as a basis for these documents and also to manage implementation, especially the organization and review of EU structural fund expenditure.

Local Government Authorities – Local Administration   Local communities (193 local government authorities in total) have the basic right to arrange the spatial management and planning of their territories, with the exception of planning control activities (planning and building permits for example, which are under direct State jurisdiction). Their principal task in connection with spatial management and planning is concern for rational, mixed, and sustainable land use, as well as the economic use of land in accordance with the principles of the high quality of living, work, recreation and public health. As such, local government authorities are required to prepare multi-year development plans for the various areas of budget expenditure. In decision-making procedures, they are responsible for the direct participation of all the involved and interested parties. Local authorities also care for and maintain the identity of the urban areas and the countryside by taking account of, and protecting, the natural and built characteristic features. Spatial planning documents at the local level include the Local Government Authority Spatial Development Strategy (which also includes Urban Development Concepts for cities and towns and Landscape Development and Protection Concepts for the countryside), the Local Government Authority Spatial Order, the Local Detailed Plan and Planning Information. The system thus allows a high level of autonomy for local government authorities, allowing them to draw up development strategies and programmes at the local level.

In addition to the spatial plan and the national development strategy, local long-term development activities are defined for the following year and the mid-term capital expenditure for the ensuing three years: Investment programs are generally prepared on the basis of an analysis of needs presented by the budget users, the situation in individual areas of activity, and the financial capabilities of the local government authorities.

Local government authority development plans and programmes are normally financed from both budgetary and non-budgetary sources. The budgetary sources are: the State budget, local government authority budgets, and the budgets of other local communities. Non-budgetary sources stem from the amortization of communal infrastructure, financial leasing, and loans. There are separate budgets for separate policies and development programmes. State budget funding is normally appropriated to specific programmes, depending on the ministry through which the funding is channelled to the local government authorities. In addition to these sources, private capital also plays a significant part in the financing of various local development projects. Other possible sources of finance for development programmes are the non-refundable grants provided by various international development organizations and EU institutions.

Urban Policy and the EU   The Spatial Management Policy of the Republic of Slovenia states that the future spatial development of Slovenia will depend on the country's needs and development opportunities as well as its incorporation into broader European integration following accession to the EU and in response to other globalization processes. This statement raises questions concerning, in particular, workforce mobility, the free flow of capital, real-estate ownership, and cross-border cooperation. An increase in pressure on the environment is expected as a result of foreign investments and the interests of private capital as well as additional pressures on space within infrastructure corridors (for transportation), which are of interest to the neighbouring countries and regions.

A Regional Development Agency has been created to act as the central organization for coordination of the implementation of development programmes. The Agency will also be responsible for establishing inter-regional links and for setting up representations in the relevant EU institutions. In accordance with the principles of subsidiarity and partnership, the Agency will act as the 'management body' for the regional development projects that apply for funding from the various EU programs and structural funds. According to the National Development Programmes, Slovenia is anticipating European Union funding of between 600 and 700 million euro for the period 2004–2006, after joining the EU. The projects already in line for finance from EU structural funds include: a public transport project intended to reduce the problems of traffic congestion in Ljubljana; a project for the modernisation of communal infrastructure intended to improve the quality of living in urban areas; and a project for the modernisation of the education system and its linkage with the commercial technological park.

National Urban Policies

Slovenian urban policy defines several important objectives, which may be summarised as follows:

- to redefine and implement the concept of polycentric development of cities and to ensure their equivalent integration into the European urban system;
- to strength sustainable development, to promote re-urbanization processes, rational land use and revitalization of degraded urban areas;
- to improve the relationship between town and countryside;
- to provide equal opportunities of access to information and knowledge.

Redefining and Implementing the Polycentric Urban System   The development of a balanced and polycentric settlement system was already defined in the long-term plan of the Republic of Slovenia in the year 1986. Since 1991, after Slovenia decided to join the European Union as an integrated member state, the efforts to achieve a better territorial cohesion have been intensified. According to the Spatial Management Policy of the Republic of Slovenia, a 'harmonious urban network development is a prerequisite for the overall development of the country, and for this reason it has to be guided in an effective way'. Following this, the functional and physical interconnection of towns and other settlements in the urban network will be reinforced, by, amongst others, the development of good public transport and telecommunication links. At the national level, the urban network comprises three centres of international significance: the capital city of Ljubljana, Maribor, and Koper (the Koper- Izola-Piran coastal city conurbation); and 5 centres of national significance: Celje, Kranj, Murska Sobota, Nova Gorica and Novo mesto. These centres are a motor for economic, social and spatial development within each of their functional regions (8 in total). Slovenia also aims to enhance its polycentric development at the regional level 'providing conditions for economic efficiency, a balanced distribution of job supply, service activities and housing, as well as caring for the quality of the environment' (Zavodnik Lamovšek 2003: 14) . The introduction of the regional planning level (the new Spatial Planning Act, 2002), the adoption of the Balanced Regional Development Act, the Regional Development Strategy and the National Housing Programme (2000–2009) are the best indicators that Slovenia is striving intensively towards sustainable regional development.

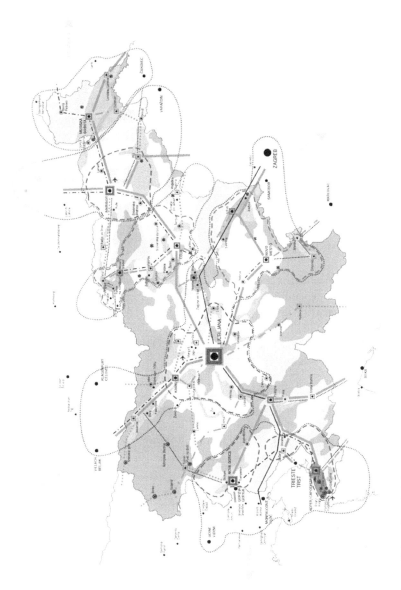

Figure 5.1   Spatial development concept of Slovenia
Source: Urban Planning Institute of the Republic of Slovenia, 2002

Strengthening Sustainable Development: the Promotion of Re-urbanization Processes, Rational Land Use and Revitalization of Degraded Urban Areas The preservation and (re)development of towns and other settlements is a high priority in the polycentric development of the settlement structure. Special attention is being focused on rational land use and the revitalisation of the existing city centres, as well as the re-urbanisation of former industrial zones, abandoned military sites and other degraded urban areas. In this respect, particular attention is being paid to high-quality planning and the protection of the environment and of the natural and cultural heritage. In the newly shaped urban patterns, a high quality of life is being striven for by mixing various urban uses of space, designing quality natural areas and through careful urban and architectural planning including a revival of environmentally- and user-friendly public transportation. An important condition for the development of urban centres is the quality of the living space, which in general, remains one of Slovenia's advantages in comparison with other European regions. The attractiveness of the urban areas and landscapes can only be enhanced by a rational allocation of land use, better quality of urban and landscape design and good architecture, in harmony with the elements of natural values and cultural heritage. These are also the main resources of the Slovenian tourist economy.

Improving the Relationship Between the Town and Countryside The national urban policy is also 'encouraging development generated by urban functions and improving the relationship between town and countryside'. The connection of towns with the countryside is one of the basic principles of sustainable spatial development in Slovenia. The new Spatial Development Act introduces measures which will gradually redirect spatial development from the spontaneous concentration of settlements around major city centres (suburban areas) to the revival of city centres on the one hand, and the preservation of remote rural areas on the other. National urban policy also seeks to make use of Slovenia's comparative advantages and protect its national identity in the European integration processes, taking into consideration its regional characteristics and the spatial diversity (e.g. the proposal for protection of 30per cent of the state territory under Natura 2000).

Developing Access to Information and Knowledge The state encourages the development of information technology and information transmission via various systems. For the purposes of the development of access to information and knowledge, Slovenia has adopted the National Strategy that lays down the developmental directions of the country in the narrower area of telecommunications, as well as in the broader context of the development of an information society (i.e., society based on knowledge). Important objectives of the National Strategy relating to the creation of urban policies are enabling access to the services of the information society to the greatest possible number of citizens, as well as training and creating new working methods and the intensive introduction of electronic services into public administration with the preparation of digital planning documents, and in local self-government operations.

The implementation of some urban polices, for example the preparation of standardised digital planning documents at the local level, has been co-financed by the Ministry of Environment, Spatial Planning and Energy, because of their importance for the future monitoring and preparation of reports on spatial/urban changes, that are due to be prepared by the municipalities every second year. At present, the national government has power to monitor and prevent changes from agricultural land into building land. In the process of reviewing planning documents, all the changes of agricultural land into other urban uses should be clarified and justified in detail, with permissions from the Ministry of Agriculture, before being approved by the Ministry of the Environment, Spatial Planning and Energy and adopted by Municipality Councils. However, policies for integration of housing construction, land-use and public transport, similar to the sustainable urban management policies are still missing!

Other Relevant Policies

There are several other policies that address, more or less directly, a variety of urban development issues, and define the measures for dealing with various urban problems. The most important among these, related to safeguarding sustainable development (Sendi et al. 2002, 2004) are as follows:

Transport Infrastructure Programme The main aim of this programme is to improve the condition of roads and to facilitate a balanced development across the country. The national road network must enable adequate access to all geographical regions and provide efficient and economical internal connections throughout the country, taking into account also those regions facing depopulation, border regions and tourism areas.

Solid Waste Management Programme Priority investment projects in municipal waste management are directed at activities for gradually decreasing the amount of waste and its possible or latent hazardous nature at the source, waste sorting and increasing the degree of the material and energy value of waste. Priority investment in industrial and energy production waste management is geared towards the reconstruction and enlargement of hazardous deposit sites in accordance with the EU standards.

Housing Construction Programme The Housing Construction Programme determines three main priorities: the implementation of the National Housing Programme (NHP); the implementation of the National Housing Savings Scheme (which offers savers favourable loan conditions for housing purposes, at the end of the determined saving period) and the implementation of the mapped out rental housing policy. Increasing the supply of building sites is of special importance. In the period 2001–2006, there are plans to construct 35,000 privately owned new housing units, 1,350 private rental and 10,100 non-profit rental housing units. The key investment priority is the construction of non-profit rental housing units in close cooperation with the Housing Fund of the Republic of Slovenia and local communities in the co-financing of construction and the provision of building sites.

Equal Opportunities and Social Cohesion Programme This program has two main objectives: to promote social cohesion in such a way as to enable socially excluded and marginalised persons to reintegrate into the labour market; and to improve the position of women in the labour market. While this concerns wide sectors of the population in Slovenia, attention is particularly focussed on those over 50 years of age, disabled persons and young people with special needs.

Employment and Life-long Learning Programme The goal of this programme is to ensure a stable growth in employment, reduce unemployment and prevent long-term unemployment. The programme is based on the existing employment policies and focuses on improving the quality of existing programmes. The programme will support the implementation of the National Education Strategy in terms of embracing the concept of life-long learning and improving accessibility to learning in the education system and vocational training. Learning has to be seen as a precondition for improving the potential for progress and transfer of basic knowledge. Employers and employees have to realise that life-long learning is a prerequisite for economic growth, competitiveness and social cohesion.

Prevention of Social Exclusion In the area of social policy, the State aims to prevent social exclusion, particularly by influencing the social position of the population in taxation, employment, and work, and through grants, housing policy, family policy, healthcare, education, and other relevant policy areas. In fact, all services are relevant, because of the high numbers of residents living in very different conditions. Some social services are the responsibility of the State, while others are the tasks of local government authorities, or under joint responsibility. Based on the Social Protection Act, the Programme of Combating Poverty and Social Exclusion, and the National Social Protection Strategy, the following services have been provided: initial social care; personal assistance; family assistance; institutional care; guidance, care and employment under special conditions; internal assistance to workers in companies, institutions, and in cooperation with other employers. Social considerations also apply to the housing field. Within the NHP, special attention is paid to the housing needs of vulnerable groups such as young families, the disabled, persons with mental problems, and the elderly.

The policies presented above show that national urban policies relate to general urban development problems of concern to the entire country and not necessarily to concrete urban problems in specific areas. Specific urban policies and programs for dealing with specific urban problems are prepared and implemented by the relevant local government authorities. It may be observed that national urban policy gives priority to the realization of the long-term desire to effect a polycentric urban system in Slovenia. The principal underlying logic is to achieve a more even settlement pattern that may lead to more balanced economic development over the entire territory. This policy also seeks to guarantee a more rational and efficient use of available resources, thereby contributing to a reduction in environmental degradation.

Spatial Development Trends and Urban Problems

The Spatial Management Policy of the Republic of Slovenia states that the future spatial development of Slovenia will depend on the country's needs and development opportunities, as well as its incorporation into broader European integrations following accession to the EU, and in response to other globalisation processes. This raises questions especially concerning workforce mobility, the free flow of capital, real estate ownership and cross-border cooperation. An increase in environmental pressure is expected as a result of foreign investments and the interests of private capital as well as additional pressure on space within infrastructure corridors (transportation), which are of interest to the neighbouring countries and regions.

Current urban changes in Slovenian cities, significant for these post-socialist cities, have been associated predominately with commercialisation and gentrification of the historic core, reurbanisation and revitalisation of some inner city areas, and residential and commercial sub-urbanisation in the outer city (Dimitrovska Andrews 2002, 2004).

Historic Core

In the historic cores, the development of offices, multipurpose commercial centres and tourist oriented facilities, by refurbishment of existing buildings, or new infill development together with gentrification promoted by the private sector and city government, are the prevailing interventions. Unfortunately, there are also 'negative' consequences of these revitalisation processes, such as a marked decline of residential uses, conflicts between the interests of commercial development and the protection of cultural heritage, traffic congestion and parking problems. However, strengthening and developing a distinct urban structure has been accepted as a key value in the preparation of a new generation of planning documents. The main aim of these documents is to set out the further restructuring (remaking) of the city with respect to the reinforcement of cultural identity, the continuity of the urban structure, and the legibility of public urban space.

The project Kapiteljski vrtovi (Capitol Gardens) has been recognised as good building practice in the revitalisation of the city centre of Ljubljana. The site (2,3 ha) is located next to the prominent Secessionist palace (formerly a printing press company) stretched from the Ljubljanica river on the north, to the remains of the built structure of the medieval suburb Poljane on the south. The project proposed the renovation of the Secessionist building for the Faculty of Law, and reconstruction on the site along the river embankment, for an office building, with a residential building closing the urban block on the west side (Bavarska street). The two buildings are connected with a shopping mall and underground garage (250 parking spaces, 80 reserved for residents). The elevations of both buildings were designed with respect to the existing surroundings.

(a)

Figure 5.2a   Kapiteljski Vrtovi

(b)

Figure 5.2b   Nove Poljane

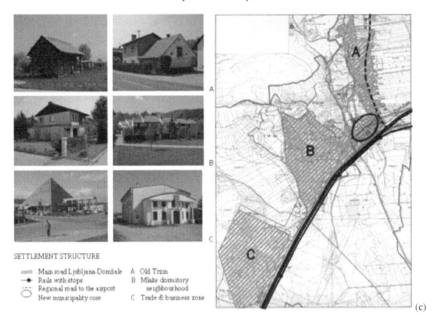

SETTLEMENT STRUCTURE

=== Main road Ljubljana-Domžale   A Old Trzin
→ Rails with stops   B Mlake dormitory
⌒ Regional road to the airport   neighbourhood
   New municipality core   C Trade & business zone

(c)

Figure 5.2c   Trzin

Inner City

The most significant urban changes in the inner city areas of Slovenian cities have been the reduction of industrial and military use and release of these zones for commercial development, shopping centres and private housing ('brown-field' development). The development of secondary business nodes established in strategic locations along major access roads, and scattered housing (re)development, located in dispersed fashion on vacant zones, have been also present in most of the larger cities in Slovenia. The problems in the restructuring of the inner city areas have been associated with increasing social polarisation of large housing estates (e.g. problems of revitalisation, maintenance and management), and infill development with no respect for the characteristic identity and morphology of established city areas. The problem of degraded areas is also present. These are in most cases previous industrial and warehouses sites, which, abandoned at various stages of the transition period, account for nearly 14per cent of the total surface area of the urban centres in Slovenia. Currently, it has been recognised that there is a need for a clear strategy for future inner city development, with marketing guidance and policies for positive planning and pro-active development constraints to achieve a better vitality and viability for the city as a whole.

The residential area Nove Poljane, built on a former barracks site (4,8 ha) by the Ljubljanica river is a representative example of new development practice in the inner city since 1990. The semicircular plan for the residential estate Nove Poljane follows the original structure of the site, characterised by the curved Dolenjska rail-tracks in the south-east. The composition of the built structure consists of four rows (three in villa-block typology and one 'mega' block) from 3 storeys in the centre to 6 storeys along the rail-track. The communal park is located in the north-western corner and accessible via two radial pedestrian paths. The residential estate Nove Poljane is a rare example of good practice in a public-private partnership, where in total 440 flats have been built, 309 units in the public rental sector including 20 reserved for the disabled.

Outer City

Residential suburbanisation, as discussed before, has been present in the functional urban region (FUR) of the largest cities in Slovenia since 1970, predominately in the form of 'satellite' extensions – dormitory neighbourhoods – of existing villages. After 1985, these processes became more extensive and were followed by industrial and commercial suburbanisation. New mixed-use zones combining commercial, light industry and housing programs have been erected in the outer cities, most of them as 'green-field' development along main motorways. The most characteristic problems of the urban changes in the outer cities are visible in the coalescence of existing traditional villages into suburban agglomerations with resultant loss of village identity; transformation and loss of identity of the cultural landscape, pollution of underground water resources due to insufficient technical infrastructure and increasing individual car traffic. Therefore, effective strategic planning and coordinated action between municipalities at the regional level has been recognised as crucial for reducing and guiding suburbanisation processes.

The development of Trzin, a small settlement 5 km north-east of Ljubljana, is one of the most representative examples of planning and building practice in the outer city. Trzin has developed from a rural village with 700 inhabitants in 1950 into a large dormitory neighbourhood with several thousand inhabitants by the 1970s, and today, is turning into a vibrant centre of business and handcraft activities – a largely self-sufficient settlement, the centre of the newly established commune of Trzin.

Planning and Management

Recent development in planning and management in Slovenian cities shows positive change towards comprehensive strategic approaches to redevelopment, with the enhancement of the image of each city as a whole, and the identity of their characteristic areas. Development Strategy Concepts have been introduced giving more detailed consideration to design issues and implementation mechanisms. Other changes include:

- Transparency of planning and management of the city to improve involvement of the general public in the decision making process.

- Greater integration of physical planning and real estate regulation in order to shape the built environment more efficiently.
- Simplification of the procedures for planning permission and better responsiveness to developer needs.
- Urban renewal oriented towards the reintroduction of vital and liveable public open spaces.

Currently, for example, the city of Ljubljana is preparing a new generation of planning documents, among them the 'Strategy for Sustainable Development' and the 'Concept for Spatial Development of the City of Ljubljana' (Cerar et al. 2002). The former proposes a City Design Strategy for the urban area as a whole (e.g. enhancement of the local context, identity and legibility of the public built and open space) and Urban Design Frameworks for characteristic urban areas (e.g. rebuilding of degraded urban sites with respect to the contextual identity of their areas) (Dimitrovska Andrews et al. 2001). For implementation of the plan, the Concept is proposing three layers of instruments for spatial development: Urban Regulation Plans, Urban Design Projects and Urban Regulatory Measures (urban land policy).

Conclusions: Slovenian Sustainable Development Perspectives

The main characteristic of Slovenian urbanization is relatively small towns, which will find it difficult to be competitive with the large urban centres of Europe. Therefore connections and collaboration between particular towns on a macro level (i.e. regions) and even micro level (i.e. urban regions of larger towns) is crucial. In view of the present regional structure of cities in Slovenia their connections could be identified by theoretical models of sustainable cities (Frey 1999: 37–69) such as: star shaped/branched cities (e.g. towns in the urban region of Ljubljana, Central region of Slovenia), linear cities (e.g. conurbation Koper-Izola-Piran, Coastal region), regional towns in a polycentric network (e.g. Sloven Gradec-Ravne-Dravograd, Koroska region) (Dimitrovska Andrews 2000). We believe that development of the highway system with links to Trans-European Networks (TEN) infrastructure (transport corridors V and X) will increase the integration of small and medium-sized Slovenia towns and stimulate the establishment of functional networks contributing to balanced regional spatial development. However, increased population mobility can be expected, thus also increased functional gravitational areas of already developed centres, especially Ljubljana.

The balanced polycentric European urban system as defined in the ESDP is based on a network of urban cooperation on the international, European, and regional levels, whereby competitiveness between cities is combined with functional complementation. In the design of the Slovenian urban system it will be of paramount necessity to establish how particular centres will be positioned in the urban network, and to define the roles of:

- Ljubljana in the European network of capital cities;
- Ljubljana and Maribor in the network of Central European cities;
- the coastal conurbation Piran-Izola-Koper in the network of Mediterranean cities;
- Koper in the network of the North Adriatic ports.

The roles of other regional centres in the Slovenia urban system will have to be redefined with respect to European connections of national and regional systems and settlements. On the European level, we may expect technological and structural-economic changes of the urban system, aligned with the integrated European market and the developing network of high speed transport. On the national/regional level, changes in the urban system may be expected, especially in the border regions and along the main communication corridors. The key issues will concern possible urban development on the national periphery for tourism and secondary housing, further dilapidation of previous industrial sites and military bases, the influence of new technologies, and the problems of suburban (urban-rural belts) and uncontrolled (unplanned) urbanization.

The guidelines for achieving polycentric urban development in Slovenia are based on a sustainability paradigm and can be summarised as follow:

- to establish a balanced urban network of centres with adequate access to urban functions and prominent regional centres as significant generators of economic, social, cultural and environmental development in their respective areas;
- to ensure harmonious spatial development in areas with common developmental characteristics, especially in geographically enclosed areas (border areas, coastal and hilly areas), protected areas, and areas threatened by natural phenomena;
- to enhance the quality of towns and other settlements as a pleasant living and working environmental through internal developmental potential (renovation of city centres, rehabilitation of degraded urban areas, renovation of old industrial and mining areas;
- to strengthen, in particular, the border areas, to increase their comparative advantages and competitiveness in a broader international environment.

However, in the absence of the regional administrative level, the state continues to be highly centralised. This creates great problems in the achievement of effective vertical integration across all levels of government important for achieving sustainable development. The initiatives at the local level are consequently often undermined by decisions of higher authorities or by lack of horizontal co-operation between different departments.

Co-operation and partnership between different levels, organisations, and interests are essential, since most problems will only be solved through co-ordinated action by a range of actors and agencies. This includes public-private partnerships, as well as the involvement of non-governmental organisations, representatives of the public and the public themselves.

Efforts to achieve the desired polycentric urban system will only succeed if the following basic requirements are satisfied:

- establishment of legislative and institutional system; the strengthening of technical and governmental services and clearly defined responsibilities and tasks at the different governmental levels (i.e. state, regional, local authorities);
- provision of financial and economic measures; funds and tax policies, promotion of public-private sector partnerships;
- provision of high quality education and training of technical personnel, and development of reliable spatial information system;
- improvement and modernization of the spatial/urban management related databases, constant monitoring of trends in urban development;
- urban management promotion and encouragement of productive public participation throughout the entire spatial management process.

The four cross-cutting themes which are essential for long term sustainability of town and cites, recognised by the 'Urban environment thematic strategy' (CEC 2004) that will need further consultation and implementation in Slovenia are: sustainable urban management, sustainable urban transport, sustainable construction and sustainable urban design.

Acknowledgements

This article is based on a research programme 'Spatial development' funded by the Ministry of Higher Education, Science and Technology of the Republic of Slovenia and the paper prepared by the Urban Planning Institute (authors: Sendi, R., Dimitrovska Andrews, K. and Černič Mali, B.) on National Urban Policies in the New EU-countries: the case of Slovenia.

References

Cerar, M. et al. (2002) *Prostorska zasnova Mestne občine Ljubljana / Concept for Spatial Development of the City of Ljubljana*, Mestna občina Ljubljana.

(CEC) Commission of the European Communities (2004), *Towards a thematic strategy on the urban environment* (unpublished draft).

Dimitrovska Andrews, K. (2000), 'Urbanisation processes in Slovenia and their effects on urban networks.' *Urbani izziv*, letnik 11, št. 1/100, pp. 117–124.

Dimitrovska Andrews, K. and Ploštajner, Z. (2000), 'Local effects of transformation processes in Slovenia.' *Informationen zur Raumentwicklung*, No. 7/8, pp. 435–449.

Dimitrovska Andrews, K. and Sendi, R. (2001), 'Large housing estates in Slovenia: a framework for renewal.' *European Journal of Housing Policy*, 1(2), August 2001, pp. 233–255.

Dimitrovska Andrews, K., Mihelič, B. and Stanič, I. (2001), 'The distinct structure of the city: The case of Ljubljana.' *Urbani izziv*, vol. 12, No. 2, pp. 117–123.

Dimitrovska Andrews, K. (2002), 'Planning in flux: changes in the spatial structure of Central and Eastern European cities: the case of Ljubljana.' *Informationen zur Raumentwicklung*, No. 11/12, pp. 693–701.

Dimitrovska Andrews, K. and Černič Mali, B. (2004), 'Slovenia: Effects of privatisation.' In: Turkington, R., van Kempen, R. and Wassenberg, F. (eds), *High-rise housing in Europe, Current trends and future prospects*. Delft: DUP Science.

Dimitrovska Andrews, K (2005/forthcoming), 'Mastering the Post – Socialist City: Impacts on Planning the Built Environment.' In: Hamilton, F.E.I., Dimitrovska Andrews, K., Pichler-Milanović, N. (eds), *Transformation of Cities in Central and Eastern Europe: Towards Globalisation*. Tokyo: United Nations University Press.

European Commission (EC) (1997): *The EU compendium of spatial planning systems and policies*, Regional policy and cohesion.

ESDP (1997), *European Spatial Development Perspectives, European Communities.*

Frey, H. (1999) *Designing the City, Towards a more sustainable urban form*. London: E & FN Spon.

Keivani, R., Parsa, A. and McGreal, S. (2001), 'Globalisation, Institution Structure and Real Estate Markets in Central European Cities.' *Urban Studies*, Vol. 38, No. 13, pp. 2457–2476.

Sendi, R., Mandič, S., Černič Mali, B., Boškić, R. and Filipović, M. (2002), *Stanovanjska politika kot generator in blažilec socialnih problemov [Housing policy as a generator and mitigator of social problems]*. Ljubljana: Urban Planning Institute.

Sendi, R., Dimitrovska Andrews, K. and Černič Mali, B. (2004), *National Urban Policies in the New EU-countries: Slovenia* (unpublished).

Spatial Planning System in Slovenia, <http://www.gov.si/Assets/Homepage-eng/>

Zavodnik Lamovšek, A. (2003), *Sustainable Spatial Development in Slovenia.* Ljubljana: Ministry of the Environment, Spatial Planning and Energy.

Chapter 6

Regional Planning in Lithuania – Experiences and Challenges

Zigmas J. Daunora and Pranciškus Juškevičius

Introduction

Lithuania is once again perceived as an integral European country, owing to its Central East and Baltic regions. Together with its neighbouring countries, upon reclaiming their self-sufficiency, it aspires to overcome economic arrears as soon as possible and to reclaim its lost pre-war position. The transitional period, which lasted longer than a decade, was characterized by political, social and economic reconstruction, while urban and spatial problems were displaced into the second schedule of state tasks. Nevertheless, the first steps were made at this time i.e. after some laws regulating urban and regional planning were adopted, the country territorial managing structure was improved and some documents reflecting the new development conditions of urban and regional physical and strategic planning were prepared. However, in the framework of the strengthening economy there is a clear necessity for a more precise determination of state regional policy provisions, more rapid changes in the legal spatial planning basis and stronger regulation of urban development in cities and the surrounding areas. By the same token, work in such fields has a strong position and progressive achievements from previous development. We would like to list here some urban planning problems in the context of EU priorities as well as achievements in Lithuanian urban system sustainable development.[1]

Fountainheads of the Planning Works

Lithuania has a distinctive system of human settlements. Under different political, geographical and economic conditions the settlements constantly experienced changes, i.e. as some of them declined, others grew and strengthened. The origin of new towns is said by many authors to coincide with the formation of the state in

---

1   The Republic of Lithuania is divided into administrative units: counties and municipalities. A municipality is a territorial administrative unit administered by the self-government bodies elected by the community. A county is a higher administrative unit, one of 10 regions of the territory whose administration is organized by the Government of the Republic of Lithuania.

the 13th century. Subsequent state legal acts (e.g. agrarian reform by Sigismundus Augustus) and privileges separated the Lithuanian settlements to a greater degree into towns, townships and villages. The network of towns having Magdeburg privileges and townships with market rights became clearer.

In the long process of town and township development there was a visible trend towards keeping optimal distances between settlements as determined by competitive economic conditions and the rise of towns as leaders in particular areas that functioned as culture and management centres. The measures of the day for legal regulation had a fair importance for the self-contained process. The reform provided in the second part of the 16th century essentially reshaped agrarian relations between estate and village, the agrarian system, forestry, and planning and management of privileged settlements (Jurginis 1975). As Lithuanian towns expanded their relations with the centres in neighbouring countries, inter-state tracts and fields of economic interaction began to be formed. The settlement network, which took hold in the late 16th century, underwent comparatively few changes, though the state and its territory later experienced extensive depredations, long occupancies and losses. Some changes and economic revivals took place in town industrial development (first part of 19th century – beginning of 20th century), especially after linking the more important towns by rail lines and with the centres in other countries.

The historically short period of independent development (1918–1940) was very important for the handling of towns and townships of the re-established state of Lithuania. The Law on Land in 1922 prescribed the methodical formation of towns and townships. Traffic problems and laying of new roads were taken into account. Many modern buildings designed by professionals who had been educated abroad (in Italy, France, Germany, etc.) began to be built in Kaunas. Professionals were trained in the re-established Lithuanian University and Technical School as well. Interest in urban planning grew because there was a need for complex decision making with regard to urban technical, social, economic, hygienic and aesthetic problems. Preparation of expensive topographical plans of towns and townships was undertaken (Salkauskis 1938).

Directive Stage of the Urban Development

The difficult Lithuanian economic and political situation, especially with Soviet industrialisation and collectivisation, inspired the heavy migration of countryside populations into cities. This migration, by its spontaneity and degree, was out of character with the formerly quiet and self-contained development of Lithuanian settlements. The unpredictable outcomes and clearly visible necessity for spatial regulation of public processes raised a great deal of concern among urban scientists, geographers and economists. Positive conditions were taken from an urban doctrine, which prioritised the development of small towns in order to stop the hypertrophied development of the bigger cities, which was prevalent in Western Europe at that time and has been catalogued in the former Soviet Union. By the early 1960s a concept had been formulated in Lithuania that expressed a scientific united system of settlements (Prof. K. Seselgis and others). The essential tenet of the concept was that the network of towns and villages would be a united system and that their problems could not be solved in isolation. One of the most important city functions was to be the provision of social and industrial services to the network of surrounding small settlements to expand its economic and cultural potential (Table 6.1).

Table 6.1 Model of the united resettlement system in Lithuania, 1964

Structural levels of the system elements	Quantitative indicators of elements of the system				
	Space units			Space unit centres	
	Number	Area (km^2)	Population (thousand)	Access (hour in travel)	Population (thousand)
I Country	1	65,000	≥ 300	-	>500
II Region	10	6,500 (average)	≥200	2.0	≥50
III District	36	1,800 (average)	≥30	0.5 – 0.75	≥10
IV Rural territorial district	250	150	≥3	0.25	≥1

Source: Seselgis 1996

By relying on the above-mentioned principles a polycentric urban system was to be developed in Lithuania. In parts of the country where powerful centres did not exist, the expansion of the functions of prospective middle-sized and small towns was proposed. Shortly (1964), this scientific concept was implemented in the first Lithuanian regional planning scheme. Contrary to the declarations of decentralized development principles in other countries, these ideas met with accord and strong results in Lithuania. Practically, from the beginning of the 1970s Lithuanian urban planning, functional and demographic development was linked with the common interests of the country. The statistical data shows that strong urbanisation took place for 40 years (1950–1990) (General Statistics Division of Lithuania 2002). During this period the urban population increased from 28.3 to 68.5 per cent. Only 31.5 per cent of the population remained living in the countryside.

Results of the spatial planning scheme were especially visible from 1970-1990, i.e. from the epoch when the urban and rural populations became equal, to the end of the intensive urbanisation process (1990), when the proportion of urban and rural populations stopped changing. During this period the growth rate of the two biggest Lithuanian cities (Vilnius and Kaunas), compared to the national average, was successfully reduced by 11 per cent, and the growth rate in 16 middle-sized towns was even increased by 36 per cent (Table 6.2).

Table 6.2 Development of Lithuania's cities and towns in 1970–1990

City/town groups in 1990	Population, ths people	Percentage	Growth rate 1970 = 100 %	Comp. with aver. growth %
Very large (2)	1,022.2	40.1	150.9	-11.1
Large (3)	482.8	18.9	157.6	+4.4
Medium (16)	522.7	20.6	198.1	+36.1
Small (94)	521.7	20.4	148.0	-14.0
115	2,549.4	100.0	Av. 162.0	

Source: Daunora 1996

The global urban problem in that century – concentration and decentralisation – was solved in favour of the latter direction. This important and prospective provision needed continuous attention and endeavour. The centralized command management heavily respected 'local' interests. Therefore, it was difficult to avoid the construction of inorganic, very large and ecologically dangerous enterprises that only aimed to tighten economic and demographic links between Lithuania and the USSR. For that reason the officially accepted strategy for small town development was even more important because it stopped the growth of the biggest cities and decreased the associated ecological and national problems, helping to avoid depopulation in the periphery zones and consolidating the role of the new centres (at first Utena, Alytus and Tauragė) in the national territorial structure. In comparison with the neighbouring Baltic countries, the signs of polycentrism in Lithuania became very clear (Table 6.3, Figure 6.1).

Table 6.3 Urban characteristics and largest cities' role in the states of the Baltic Region (1990)

State	Population, million	Density, people/sq.km	% of urban population	Population of Capital		Share of next 4 (2-5) cities in the country, %
				m. people	% of the country	
Lithuania	3.75	57.5	68	0.60	16	25
Latvia	2.66	41.1	70	0.90	34	14
Estonia	1.53	34.6	58	0.45	30	19
Finland	5.05	14.8	62	0.51	10	13
Sweden	8.59	21.1	86	1.04	12	10
Denmark	5.16	119.8	85	1.34	26	10
Norway	4.24	13.1	74	0.46	11	12

Source: Baltic Institute 1994

Regularity of Dispersion of the Integration Processes

Later investigations toward developing a theory of a united settlement system (Daunora 1984) showed that items of the spatial structure do not cover all planning problems. The rationality of the regional framework depends on how the structure of its plan expresses the integral, single-functioning nature of its composite parts. Identification of the planning structure and its subsequent elaboration is no less important than the formation of the regional administrative structure. By investigating peculiarities in the integration processes it was determined that in a small compact country there exist very different settlement integration conditions in the country's centre and periphery. It happens consequently that the central part is affected by several big cities. Favourable conditions promote public processes and the urban structure is formed in the central part (the so called 'urban heart' or 'the urban stem'), which consolidates the country's territory. The structure reflects the self-contained trend of concentration of public processes. It provides the rationality and compactness for the spatial links. With them the role of the individual centres grows with respect to cultural, social and economic services in the less integrated frontier areas.

Identification of the spatial distribution for the urbanisation process and formation of the country space structure clearly indicate the urban stem in Lithuania. It is fixed by the 4 biggest cities – Vilnius, Kaunas, Siauliai and Klaipeda, important interior and exterior communication elements, peculiarities of the spatial dispersion of the natural and demographic potentials, etc. Acting together, these elements increase the specialisation of the various country parts. The result is a positive polarisation into intensively urbanised and less urbanised territories for recreation and agriculture and forestry (Figures 6.1 and 6.2).

At present, maintaining both the proportions and specifications of urban zones and rural areas (agricultural land, forests and other territories) has become an important condition for a sustainable structure. In order to achieve this, a coherent long-term policy is necessary; therefore, new proposals should support the achievements of previous stages, particularly with regard to social and economic activity, and should share activities such as natural and cultural preservation among different regions of the country.

Among actual tasks for further development, the problem of the proportions of the urban and rural population is under consideration. For countries with a moderate population density (there are 53.5 inhabitants/sq. km in Lithuania), it is important to preserve a dense network of small townships, national parks, agricultural land and forests, and services for environmental protection, and to expand recreation and tourism in rural areas. The social aspect should be taken into account as well – the human right to choose a living place and a way of life. The rural settlements – townships, villages and farms – expand such possibilities. Therefore, these forms of settlement and their exclusive features should be protected. A directive bias in favour of rural inhabitants is improper today. Migration processes are regulated by employment opportunities and the possibility to expand certain activities. EU support opens new prospects for rural tourism, natural protection and ecological agriculture with the aim of preserving the dispersive resettlement forms and regional cultural traditions. Meanwhile, the demographic prospects for the growth of Lithuanian cities, at least up to 2015, are not favourable. The demographic situation is declining even more due to increasing migration of the younger population into more developed countries. Therefore, our regional policy should be predicated on real possibilities and clearly determined priorities.

Further Benchmarks of Regional Development

The principles of regional development consolidated in the first planning scheme for the Lithuanian territory (1964–1980) were the primary influence not only on the settlement system but also on the entire country's development. During the elaboration of the scheme, specialized schemes were prepared (nature preservation, recreation, industry decentralisation, development of transport systems, etc) that have not lost their importance even today. However, even with directives limiting the growth of the main cities (Vilnius and Kaunas) and expanding a group of smaller ones, some very clear concentration attributes of geographical, cultural and industrial potentials remained. After 1990, due to the transition to a market economy, many large-scale industrial enterprises collapsed; meanwhile, new investments were attracted by the biggest cities. Therefore, differences in the quality of life in various parts of the country dramatically increased, endangering the stable development of the country. For instance, Vilnius county took 65.5 per cent of total direct foreign investment in 2002, while all five to-be-developed regions took only 4.2 per cent. The disproportions exist in the distribution of GDP as well: Vilnius, Kaunas, Klaipeda and Siauliai counties together had 73.7 er cent of GDP, while the remaining six counties had only 26.3 per cent.

A further development of new regional centres is actually off the table because: (a) there is no expectation of collecting the monstrous resources at or near the pace we had a few decades ago; (b) investment procedures, especially private ones, are not under direct management nor can they be accurately predicted.

The retrievals of new benchmarks should be accompanied by the understanding that an uncertainty exists in market conditions, and at the same time planning can deny neither the composed basis nor the processes that are under formation or have formed and are developing further under the influence of this basis.

The planning basis consists of:

a) The geopolitical situation in the Republic of Lithuania; general demographic, social and economic national conditions; environmental quality; land use; urban system (hierarchical centres, their zones of influence);
b) Existing information. It allows making only short-term extrapolations of some processes. The course of other processes may be charted by repeating general trends in other countries with a high expectation of repeats in Lithuania.

Planning should be based on:

a) Trends of internalisation of the national economy and influence of the world market as well as interests of the interior market and its possibilities;
b) Principles of sustainable development;
c) Inertia of country development, uncertainty of further evolution and possibility for only indirect regulation of processes;
d) Separate programs should be prepared for the country or its parts or programs for problematic areas as well as comprehensive and special plans. The model of spatial development shall be simple, easily understandable and pragmatic.

Trends in regional development are:

a) Limited increase of the population of Lithuania. The reality is only the separation of the population between urban and rural areas, small and big cities, districts and regions. Separation processes (directions of migration, its size, rate, etc.) will be conditioned by the employment supply, receivable incomes, life conditions, services' quality and other warranties, their essential differences in regions, districts, cities and villages as well as other impartial changes of the population employment structure;

b) Employment turnover: employment in the first sector (mining works, fisheries, agriculture and forestry) will decline. The pace of the decline depends on the size of the agriculture and forestry market and peculiarities of the production needs, pace of land reform, productivity of agriculture and forestry, etc. It will affect regions and districts where natural conditions determine an expectation of small or medium benefit from the agricultural sector; employment in the second sector (production, energy, water supply economy, construction) will grow at a moderate pace. It will depend on more modern technology, the restructuring of production, etc; further growth of employment in the third sector (trade, transportation, communications, services, and finances, insurance, etc.);

c) Predicted changes in land use: protective areas will not decrease; demand for protective areas will increase, especially where recreational activity may serve as an economic basis. This need is connected with 'overlapping' activities (agro-tourism, urban-tourism); demand for transport and engineering infrastructure will increase; agricultural land areas will decrease. Forestry areas will increase, replacing unusable agricultural land; demand for areas for industrial production will remain undefined. This is connected with industry restructuring and the possibility for more effective use of existing zones and districts devoted to industry (production); urbanised areas will increase, at first on the outskirts of main cities with favourable possibilities for investment. Activity restrictions will increase due to requirements for environmental protection.

A Differentiated Model of Regional Development

Due to very different trends in existing development in a small country, each region may be called 'unique'. Conceptually the development trends may be outlined by the three zones principle (Juškevičius 1999): active evolution, regressing evolution, 'buffer' (overlapping) zones.

- The active (relatively active) areas are formed by larger cities – i.e. centres of industry, science, culture and services, and other cities with inherited potential for industry, railway lines and main motorways.
- Regressing areas are formed by districts where the main economic basis is agriculture, forestry and recreation. Furthermore, geographical and communication isolation from the biggest cities is typical for them.
- 'Buffer' areas, by their system of settlements and the nature of their activity, are similar to regressing areas. However, due to a more favourable geographical location and less isolation (somewhere offering more favourable conditions for agriculture), they may be marked by more auspicious processes compared with regressing evolutional areas.

Active (relatively active) areas, from a macroeconomic standpoint, are of primary importance and condition the entire national development. There are no arguments that can deny their importance, and the possibility of other territories forming to replace the first ones does not exist. Meanwhile, the demographic, social and economic processes in the regressing territories are the outcome of the development of historically formed natural and demographic potential, which might be changed only in the case of new very significant development factors. There is a great expectation that the evolution of 'buffer' areas can turn in different directions, i.e. some of them may integrate into active (relatively active) territories and others may become regressing (Figure 6.3).

The existence of problematic areas is legitimised in the new (2002) Comprehensive Plan of the Republic of Lithuania.[2] Hence, there are areas legitimised by law where the market process and municipality endeavours cannot solve the main problems of development without state support. Therein is the legitimised existence of active social, economic, cultural, and other activity zones as well as their status as zones for microeconomic stabilization with sustaining conditions for further favourable development.

A balancing principle is applied to problematic zones and zones of active development. This principle means that the state legal basis not only does not allow backtracking in social, environmental and health care, finance, economic, quality of life and other areas but also promotes an approach of both the active and regressing zones. However, the shifting of existing negative processes into the proper direction is possible only by compounding amendments into the legal basis. They should be auspicious for any investment and initiatives (state, municipalities, private persons and structures). The structure of the regulative mechanism should include grants (for example, to agriculture), subsidies, technical assistance and other help, funding support, etc.

These items are declared in the new Comprehensive Plan of the Lithuanian Territory together with a strategy in the opposite direction, i.e. 'the formation of a centre of the European level as an urban dipolis Vilnius – Kaunas by connecting the currently existing potentials of these metropolis centres'. In the prepared Vilnius City Strategic Plan (2003) the main arguments are (Vilnius City Municipal Enterprise 2003):

- Lithuania does not have a metropolis at the 'Euro-City' level, which means a danger that only a secondary role may be attributed to the state in developing the economy, attracting investment and consolidating versatile partnership relations with EU countries.
- Therefore all powers shall be focused in Lithuania on the establishment of a powerful metropolitan centre, i.e. an urban formation with capital status, population of a million, and international recognition, etc.

---

2   Comprehensive plan of the territory of the Lithuania (2002), published in: special volume of the monthly Journal Build and Architecture (Statyba ir architektūra), October 2002: 3–39.

The question arises, whether Lithuania could play a first-priority role in attracting investment and strengthening links with EU countries by using old methods, which are not tolerated in Europe (ECE/UN 1996). There is a lack of more serious reasoning. A collage of phenomena having very little to do with the Lithuanian situation is applied for the seemingly scientific 'dipole' substantiation. As often as not the presented outcomes are not justified by the facts (Bardauskiene and Vanagas 2003; Vanagas and Staniunas 2000). In our opinion the common Vilnius and Kaunas urban development would highlight even more the presently noticeable disproportions in spatial development.

There is no doubt that the biggest cities play an important role in their countries' cultural and economic development and in international integration. However, Vilnius and Kaunas would be more noticeable if they did not interact amongst themselves but with recognized centres in other European countries. Only representing their own (Vilnius or Kaunas) municipalities but not a certain non-formal compound – a town-village extending more than 100 km – would provide more possibilities to join into the European city network. Proposals to form an urban structure of a common city are unconvincing, especially today with the prevalence of talk on the advantages of compact cities, the expanding necessity of limiting urbanisation along transport arteries, and evolving international co-operation in border areas (Hoggett 1999, Juškevičius 2003, Pichler and Milanovich 1997, Rogers and Power 2000).

The 'dipole' problem does not exist in Lithuania de facto. The two biggest cities Vilnius and Kaunas have different functions and ranks in the national urban system. The directions of their territorial development are determined by local conditions. The big cities successfully grow, defeating smaller ones with their attractive power. Only sustained development requires local attempts and state regulation. Therefore, the state's documented goal of joining the potentials of the two biggest cities is equivalent to monopole structure development in the Lithuanian urban system.

A 'dipole' formation would shift the existing polycentric system of Lithuanian cities in the other concentration direction. It could be assessed as a step backward instead of progress. The outcome would be one in which we would not have a clearer and more sequential policy of regional development that could be substantiated by scientific analysis of existing problems and real possibilities and priorities of modern development. The discipline was brought into an obvious crisis due to inadequate attention from state institutions to the problem of regional development. The scientific nature of the problem was replaced by speculative declarations (Daunora et al. 2004).

There is no available basis for the formation of a dipolar urban structure because the development zones in the cities' informal regions do not overlap, the links are weak and there are no prospects for a linear urban structure formation. Thus, the Lithuanian urban structure development should be linked not with enlarging the dimensions of its physical elements but with the interior and exterior development of high speed transportation and information technologies, which should form the basis for the cities' co-operation and competition not only within the country space but also outside its borders (Juškevičius 2004).

In our opinion the promotion of specialized development of the biggest cities should be considered a more important task for the state spatial strategy. The present phase of the 'transitional' economy provides favourable opportunities for shifting the functional orientation of the biggest cities, which under the conditions of the directive economy had lost their individuality, in a more desirable direction. Specific governmental co-ordination and support (investment, development of self-governance, etc.) should give impetus to finding the basis for their cultural and economic revival more quickly. Specialized development should mean increasing every city's role in the country and beyond its borders and wider market space. It would entail the revival of embodied traditions by choosing specific directions for economic activity. Even now some differentiation for the biggest cities is provided by external transport conditions, terminals that are foreseen to be built, directions of scientific potential, cultural traditions, etc. Different forms of specialized development, i.e. social services, business, recasting of raw materials, etc. could successfully expand the smallest county centres without expanding the big industrial towns. There are relevant issues, for example, the perfection of the state territory management structure. The perfection of the territory management structure performed in 1994 was progressive and expected. However, the 'rough' trimming – without changing the inherited administrative district borders – has various defects, is paradoxical and should be repaired.

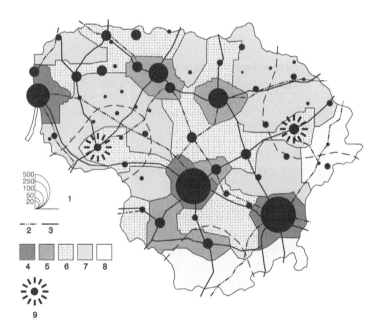

Figure 6.1a   Territorial development of the urbanization process (according to the complex intensity of social, economic and demographic processes). 1: size of city/town, in ths. inhabitants; 2, 3: main connections of the communications system; 4, 5, 6, 7, 8: concentration of various social processes in a sq. km; 9: areas where urbanization should be developed (1980).

Source: Z.J. Daunora / P. Juškevičius

Figure 6.1b   Prospective urban structure of Lithuania

> 1, 2, 3: axes, cores and branches of the urban stem; 4: rest and recreational territories; 5: agricultural and forest areas; 6: other important towns – district centres.

Source: Z.J. Daunora / P. Juškevičius

Formation of the New Lithuanian Urban Planning System

There was not a single legal act on territorial planning with regard to the formation of a natural and anthropogenic environment in Lithuania during the entire Soviet era.   Instead of a law there were acting directives, norms and rules, various recommendations, methodologies, guidance, etc. issued by the central authority. These   comprised a complete system of normative planning documents, whose structure and content coincided with the state central planning and managing principles, contemporary political provisions and understanding of the efficiency of economic, social and environmental protection. By supplementing this methodology with attributes of confidentiality and determinability it is possible to obtain a fundamental picture of the contemporary planning process.

The Seimas (Parliament of the Republic of Lithuania) embraced the first Law on Territorial Planning in 1995. It should be employed for the new planning system formation.

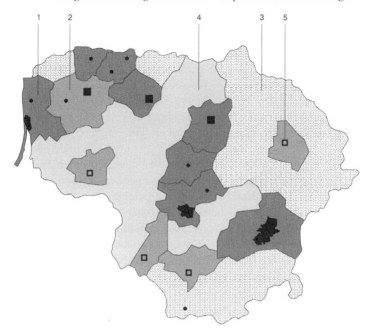

Figure 6.1c   General description of the Lithuanian land development
 1: areas of active development; 2: areas of relatively active development;
 3: areas of regressing development; 4: "buffer" areas; 5: Utena, Tauragė,
 Marijampolė and Alytus districts in "buffer" and regressing areas with
 relatively active development (2000).
Source: Statyba ir architektūra 2002

The law did not foresee any stringent overly vertical order of priorities for the preparation of comprehensive plans (i.e. preparation of the country comprehensive plan at first) or otherwise (preparation of a municipality's comprehensive plans at first), nor any terms for comprehensive plans preparation. In our opinion, it was a guiding principle to avoid a formal treatment of the law. Planning was executed by interested municipalities that could see a real benefit in it. Therefore, in order to coordinate neighbour plans there was no need to wait for the preparation of comprehensive plans in other municipalities, counties or the country.

The original feature of the first Law on Territorial Planning was the legitimisation of permanent planning. Permanent planning had to provide guarantees for so called operative planning, i.e. monitoring the development, analysis, assessment and preparation of amendments to the comprehensive plan. It encouraged municipalities to establish their own planning enterprises (for instance, the enterprise 'Vilniaus planas', which became something like a large skilled planning centre). The appearance of these enterprises was very important because they replaced huge decaying urban planning institutes. Small municipalities are not able to establish and maintain similar enterprises; therefore they are served by offices in county centres.

The new Law on Territorial Planning included the legitimisation of planning publicity and extensive opportunities for public participation in various planning procedures, and obtaining information on prepared decisions.

The Law on Territorial Planning as of 1995 reflected the spirit of transition to a market economy as well as some uncertainty about the rate of land privatisation, privatisation in general, investment opportunities, changes in both the way of life and value system, etc. Hence it could not be seen as a final and fixed document.

Comparative inexperience in the preparation of the new comprehensive plans is the reason for their marginal content: everything from general theoretical reasoning to concrete techniques and from detailed plans to things approaching to management tasks had a special quality. The procedures for their preparation were complicated and time consuming. Preparation of the country comprehensive plan took 6 years, and preparation of the city comprehensive plans took up to 4 years.

Recently the comprehensive planning pretenses have been replaced by strategic urban planning, with its admirable simplicity, cheapness and rapidity. It is expected that strategic planning will find an acceptable place in the system of territorial planning. But there is a fear that the emerging fascination will lead to a repetition of fashionable templates in all cities. Their value particularly declines when strategic plans are not coordinated with the comprehensive plans.

Detailed planning is mostly developed due to its relation with economic and production activity, formation of land plots, territory handling, changes in land use, etc. The detailed plan will determine the type of land use, building height limitations, density of built-up areas, intensity, building up lines, conditions for engineering network connections, services and other things related to cultural heritage and architectural requirements, etc.

The procedures for the preparation of detailed plans, co-ordination, approval and other procedures with objective consideration are complicated. There are doubts about the necessity of detailed plans for small objects. It should be assumed that detailed planning will be improved because of the common opinion that territorial planning is 'upside-down', i.e. instead of substantiated, rational requirements for the builder it became a process of figuring in advance the determined conditions and the pledge of the state, municipality or other institutions that issued the conditions. Regrettably, it moved away from tasks with regard to urban space planning and generally devalued urban planning in terms of science, creation and praxis.

Assessment of planning quality, validity of decisions and acceptability as well as relation to the sphere of the Law on Construction are the weakest links in the planning.

Plans prepared by professional planners must pass some control stages:

- Co-ordination, i.e. control of officials: Are planning conditions fulfilled? Are there no contradictions with active documents relating to various ranks of territorial planning;

- Public consideration, i.e. the stage when received proposals and contradictions may return the prepared plans to the co-ordination phase;
- Inspection, i.e. when an institution which has the right to inspect state territorial planning decides if all the requirements of the legal acts are fulfilled;
- Approval – the right to approve municipality and state comprehensive plans is delegated to politicians – members of municipality councils and members of the Seimas. For other plans the right is delegated to government and state institutions. The content of the approved plan becomes mandatory.

In practice, some urban comprehensive plan decisions can be treated as recommendations without any harm to the strategic principles of development. In this case the document would evolve into a flexible document reacting to rapidly changing conditions, opportunities and interests. Such ideas have been published in scientific publications; however, they have not been implemented. Hence, the last 15 years have been marked by attempts to find an efficient urban planning pattern suitable for the prevailing market economy conditions. The present situation is contradictory: planners have gradually obtained planning experience that partially coincides with their achieved level of market relations; meanwhile, private structures have achieved sufficiently high rank and the power to question and influence decisions on comprehensive plans.

Private structures have taken the initiative for urban development:

- They have established a system of massive trade, relaxation and service centres in selected places, which outpace the network of small shops and services enterprises. Further, they promote trips by cars, concentrating their flow and prolonging the trips;
- Present housing development typically consists of mono-functional, small and scattered (within the city periphery and outskirts) residential districts and housing groups without any social or other infrastructure, public spaces, and land reserves for the above-mentioned or other needs. Therefore, land use is inefficient and deep problems exist for both the technical infrastructure and the establishment and maintenance of public transport systems. In reality, these problems should be assessed when addressing the territorial segregation of wealthier people.
- The additional multiplexing of existing densely built-up territories is occurring rapidly. This multiplexing may be perceived as a financial and economic benefit. However, from another perspective (transport, healthy residential environment, psychological climate, etc.), it might be unacceptable, or in the best case should be considered in light of the actual circumstances and environment. The additional multiplexing of built-up areas violates the capacity balance in public transport lines by providing qualitative services.

The discrepancy between urban planning and development is governed by:

- weak land use management, which is not adapted to urbanised territories. The worst management features are: an inability to sustain the status quo where it is necessary (for instance, in reserved land plots for public spaces), and vice versa in other places, i.e. an operative transformation of a plot's borders for the structure of large spaces with later planning and reclamation under normal management. Practically it means that implementing the decisions of the comprehensive plan is enormously complicated and, roughly, the expectance of the implementation approaches $p=0.5$;
- an inefficient mechanism for co-ordinating the objective interests of both the common city and private ones, and implementing planning decisions;
- cumbersome planning procedures; unreasonable overall obligation of planning decisions; overly high specification of the comprehensive plans without relation to strategic planning objectives; insufficient legitimacy of planning decisions and relation to the investment process.

The efforts to improve the planning system continue: the second Law on Territorial Planning has been in effect since May 2004. As a first impression, it seems as if the law has moved towards asseveration and more complicated planning procedures; it is unclear if the evolutionary nature of the planning system will be sustained or its positive achievements will be denied. Directive provisions emerged in the Law: the obligation to prepare comprehensive plans in a hierarchical manner (from the state comprehensive plan to the counties; afterward, the municipalities and even later (authors' note) the plots' detailed plans; there is a tentative order that all comprehensive plans are to be prepared by late 2007.

Conclusions

Summarizing the modern development of the Lithuanian urban system and the problems of city and regional planning, the outcomes may be stated as follows:

1. A few qualitatively different stages are visible in Lithuanian regional development: (a) regeneration of the economy, which was destroyed during the war and during the implementation of the Soviet directive economy (1945–1963); (b) decentralized urbanisation in the successful realisation of the first Lithuanian territorial planning scheme (1964–1980); (c) further inertial realisation of the first scheme's principles with unsuccessful experiments in changing the strategy (1981–1989); (d) transformation into a market economy, stagnation of regional and urban planning and popularisation of inadequate ideas (1990–2004).
2. Necessary steps in Lithuanian urban and regional planning are: (a) respecting the results of the decentralized urban network which meets EU specifications

and has been achieved in Lithuania; (b) harmonisation of interstate integration programs with the interests and possibilities of the national economy and culture, in the search for a balanced interface between interior and exterior factors; (c) substantiation of urban and regional development ideas by scientific argument, existing problems and real state demographic and material expectations, primary attention and state resources should be devoted to decomposition of regressing areas.

3. There is no basis for the formation of a dipolis in the Lithuanian urban structure. The potential development zones in the country's biggest cities and their non-formal regions do not overlap, the internal links are weak and there are no indications that a linear urban network can be formed. The development of high speed transport and information technologies as well as integration processes should serve the evaluating aspects of the non-decreasing competition between the cities.

4. The regulation of city and regional planning and development is closely connected with the problem of establishment of the legal basis. More attention should be paid by the responsible state institutions to the above-mentioned items by including the scientific potential of the regional planning and professional planners into the country's planning process.

References

Baltic Institute (1994), *VASAB 2010, Towards a Framework for Spatial Development in the Baltic Sea Region.* Karlskrona: The Baltic Institute.

Bardauskienė, D. and Vanagas, J. (2003), 'Sources for the Strategic Plan of Vilnius and Kaunas Dipolis', *Research Journal Town Planning and Architecture (Urbanistika ir architektūra).* Vol. XXVII, 4, Vilnius: Vilnius Gediminas Technical University, pp. 170–177.

Daunora, Z.J. (1984), *Principals of the Urban Structures Optimization for the Lithuanian Local Settlement System.* Moscow: CNIIPGR.

Daunora, Z.J. (1996), 'The Life without Strategy.' *Lithuanian Science (Lietuvos mokslas). Research Journal of Lithuanian Science Academy.* Vol. 10, Vilnius, pp. 31–58.

Daunora, Z.J., Juškevičius, P., Vysniunas, A. (2004), 'Do we have State Regional Policy, or why Lithuania needs two united cities?' *Domains of Culture (Kultūros barai). The monthly Journal of Culture and Art.* Vol. 11 (480), Vilnius, pp. 17–21.

ECE/UN (1996), *Guidelines on Sustainable human settlements planning and management.* New York and Geneva: ECE, UN.

General Statistics Division of Lithuania (2002), *Counties of Lithuania (2002). Economic and social development.* Vilnius: General Statistics Division of Lithuania.

Hoggett, P. (1999), 'The City and the Life Force.' In: L. Nyström and C. Fudgge (eds),

City and Culture – Cultural Processes and Urban Sustainability. Stockholm: The Swedish Urban Environment Council, pp. 388–402.

Jurginis, J. (1975), 'Valakų reformos reikšmė Lietuvos miestams.' *Lietuvos TSR MA darbai, A serija. –* 3(52), pp. 75–87.

Juškevičius, P. (1999), 'The Concept of the Country's Spatial Development.' *Research Journal Town Planning and Architecture (Urbanistika ir architektūra).* Vol. XXIII, 2, Vilnius: Vilnius Gediminas Technical University, pp. 49–55.

Juškevičius, P. (2003), *Harmonisation of City and their Transport Systems Development. Summary of research report presented for habilitation.* Vilnius: Vilnius Technical University.

Juškevičius, P. (2004), 'Bipolar of Vilnius and Kaunas in the Urban network.' *Research Journal Town Planning and Architecture (Urbanistika ir architektūra).* Vol. XXVIII, 4, Vilnius: Vilnius Gediminas Technical University, pp. 151–157.

Pichler-Milanovich, N. (1997), 'The Role of Baltic Cities in the European Urban System: Forgotten Cities of Important Regional Actors?' In: M. Aberg and M. Peterson (eds), *Baltic Cities. Perspectives on Urban and Regional Change in the Baltic Sea Area*. Nordic Academic Press, pp. 16–42.

PLANCO Consulting GmbH (2001), *VASAB 2010, Vision and Strategies around the Baltic Sea 2010 Plus. Spatial Development Action Programme*. Essen: PLANCO Consulting GmbH.

Rogers, R., Power, A. (2000), *Cities for a Small Country*. London: Faber and Faber Limited.

Salkauskis, A. (1938), *Architektūros ir urbanizacijos pažanga atgijusioje Lietuvoje – Lietuva 1918–1938*, Kaunas: Spaudos fondas, pp. 259–263.

Seselgis, K. (1996), 'Development of Regional Planning in Lithuania.' *Research Journal Town Planning and Architecture (Urbanistika ir architektūra)*, Vol. 1(21), Vilnius: Vilnius Gediminas Techical University, pp. 4–18.

Vanagas, J. and Staniunas, E. (2000), *The Lithuanian Urban System/National Urban Systems in the Baltic Sea Region.* Seminar report, Vilnius, pp. 147–167.

Vilnius City Municipal Enterprise (2003), 'Vilniaus Planas.' Vilnius City Strategic Plan 2002–2011, Vilnius.

Chapter 7

Trends and Problems of Contemporary Urbanization Processes in Poland

Piotr Lorens

Introduction

This article is intended to summarise current trends and problems in urbanization in Poland – a country of 38 million inhabitants, located in the central part of Europe, just to the east to Germany. As the country was partitioned for over a century between Russia, Prussia and Austria, and – after World War II – was subject to socialist urbanization processes, its current tendencies and problems are slightly different to those in other European countries. At the same time, in the enlarged European Union, Poland is one of the largest nations and it suffers from many problems. The same relates to its cities and towns, as nowadays they have to adapt to globalization processes.

In order to provide a clear picture of the situation in Poland, a number of different elements have to be discussed, and it is necessary to mention all the issues involved in a discussion of the problem. Therefore, this short contribution is not able to cover all the necessary elements, but – in the author's intention – it is supposed to provide a general perspective to the problem. At the same time, it was intended to discuss the background of the on-going processes, such as land ownership and planning system changes as well as the changing roles of the key urban areas in Poland. Only with this knowledge is one able to understand the current tendencies and problems of urbanization in this country.

The Situation of Polish Cities and Municipalities after the Political and Economic Transformation of the 1990s

This chapter includes a discussion on the role and spatial distribution of cities in Poland as well as on on-going changes in land ownership structure and administrative and planning systems. This overview allows for an understanding of the possible area of contemporary changes in Polish cities, which are slightly different from most other European countries.

One has to remember that the urban system in Poland during the last fifteen years was facing major changes, which were deeply rooted in both the history of the country and the needs of modernization. The so-called 'socialist city model' (adapted during the post-war communist regime) was not able to meet the requirements of the new economic order (Węcławowicz, 2003). Therefore, it had to evolve and adapt its structure to the demands of the market economy and be transformed into a 'capitalist city'. As a result, we can observe the emerging model of the 'post-socialist capitalist city', with problems characteristic of both systems, and with the domination of market forces changing the physical structure of the urban organisms. Unlike in the case of the majority of other European countries, all this is not accompanied by any major effort of the central government, and very limited attention is paid by the newly-created regional local governments.

The Role and Spatial Distribution of Cities in Poland

Almost all cities in Poland have a very long history, in many cases reaching back a thousand years. This means they were usually founded at the beginning of the second millennium, along with the incorporation of Western European models of urbanization. Also, the adoption of Christianity had a tremendous impact on the development of the urban culture. But for many centuries Polish cities served as administrative, trade and military centers, and – unlike in many Western European countries – they did not serve as centers of developing industries or crafts. In accordance with the feudal system in Poland, where one could find a large class of gentry located in the countryside estates and administrating their large farm and forest complexes (along with the peasants belonging to them), the cities did not have a major role to play. This was also true in the areas belonging to the Order of Teutonic Knights (located in the later Prussia), where cites were developed as centers of administrative control over neighboring areas. And only a very few of them acted as centers of international trade – this related to such cities as Gdansk or Krakow. But e.g. the present capital of Poland, Warsaw, was regarded for many centuries as a little city located somewhere in the countryside, and gained importance only as the location of the Royal Court. Only thanks to this factor did the city really develop as an important element of the settlement structure in Poland (Ostrowski, 2001).

Thanks to all this, in pre-industrial Poland one could find only a very few large cities, populated usually by immigrants from other countries. Both Krakow and Gdansk are good examples of this tendency. The former – the historic capital of Poland –housed the Royal Court and many institutions associated with it. Besides them, many of the inhabitants were of German origin. These were usually craftsmen, bankers and merchants, serving the King and his noblemen. A somewhat different case could be found in Gdansk, which served as the main center of Polish foreign trade. Therefore, it was populated by merchants and ship owners and operators of German, Dutch, British and Scottish origins, with a relatively small native Polish population (Cieślak, Biernat, 1975).

Unlike many other European countries, Poland was not an important center of industrialization in the 19th century. The country was partitioned between Russia, Prussia and Austria, and none of these powers was interested in developing industry in their far-away provinces, which – besides their geographical location – were considered to be 'border areas', with a high probability of being destroyed in the case of military conflict. This was changed by massive industrialization in the inter-war period and the so-called 'socialist industrialization' immediately after the Second World War, but neither factor was able to change the spatial distribution of cities. Only a very few 'capitalist' cities were founded, and one should mention here the two most important examples – Katowice (developed as the center of mining and heavy industry in the Upper Silesia region, which was founded due to the discovery of coal resources in this area) and Lodz (which became the center of the wool and cotton processing industries, and its location was convenient for both German and Russian merchants). And in the 20th century there was only one newly-founded city – Gdynia – which was intended as the main Polish port in the inter-war period (Sołtysik, 1993). And this decision was made mainly due to political reasons.

As the borders of Poland evolved and changed several times in the last century, the current area of this country is composed of three different lands:

- historic Polish areas, which are located in the central and eastern parts of the country, with a domination of medium and small-scale cities. The only big centers are Warsaw, Krakow, Poznan, Gdansk and Lodz;
- areas of former Prussia, where one cannot find – except for Olsztyn – any big or medium-sized city, and almost none of them serves as an industrial and trade center;
- areas of Lower Silesia and Pomerania (formerly in Germany), with such big centers as Wroclaw and Szczecin, and with a very limited number of medium-sized cities.

Within this framework, there is only one industrial conurbation – the Upper Silesian region. All other cities are isolated islands of development, with little impact on the surrounding regions.

Land Ownership Structure and Its On-going Changes

Contemporary Poland consists of areas belonging to the Polish state before the war and of lands formerly belonging to Germany. This has a serious impact on the current land ownership structure. Also, one should mention the special situation of Warsaw, where all city areas were nationalised immediately after the war. As a result, within the borders of one country it is possible to find three major models of land ownership and the problems associated with them:

- areas that belonged to Poland before World War II – within them the majority of land is in private hands and municipal ownership can be very limited. This situation is typical of Western European cities;
- areas that were incorporated into the borders of Poland as the effect of World War II –before which they belonged to Germany – within them the majority of land is in the hands of municipalities and the State Treasury, and since

1989 is being gradually privatised. This situation is characteristic of 'socialist' cities. And in many cases the land ownership pattern is unclear due to post-war nationalization efforts;

- areas of the city of Warsaw, which were nationalised after the war and after 1989 are also gradually being reprivatised.

In the case of all the above-mentioned types, the situation started to change after 1989, where free market processes started to change the land ownership pattern. In some cases, this led to the privatization of municipal areas, which helped to re-establish the land market and allowed development of the real estate development processes (Kucharska-Stasiak, 1999). Thanks to that in one or two decades one can expect that all the area of Poland will be more or less in private hands, and there will be no major differences between regions as regards the amount of land in public hands.

The models described above – still present in the Polish reality – heavily influence the urbanization processes. Municipalities usually are not eager to develop their own housing projects (due to a lack of money) and the only way they can sell the land to private developers is through the bidding procedure (Wojciechowski, 1999). This brings a lot of problems and developers are often not ready to participate in this process. Therefore, they prefer to purchase private land, usually located on the outskirts of cities. This is more attractive than municipal land as it is more easily purchased, with a predictable price and cheaper than in the inner cities. At the same time, cities tend to sell the municipally-owned empty plots, ready for almost immediate development. There was a large number of them in Polish cities, thanks to the not market-oriented mode of socialist urbanization. But most of the most attractive plots have already been transferred to private owners and only a few developers are interested in purchasing for high prices (the effect of the bidding procedures) plots which are built up with run-down industrial installations and which need a costly and complicated revitalization process.

Changes in the Planning System

The urbanization process in Poland is also heavily dependent on the planning system, which has changed a few times in the last thirty years. At the beginning of the economic transformation – in the early 1990s – the municipalities had to deal with planning documents originating from the socialist times. They usually included the 'large-scale growth' model of urbanization, which showed no respect for environmental or economic issues. The most important element in them – agricultural areas slated for building up – was associated with providing new areas for construction of the new housing estates. What was also interesting for this type of plans was the fact that no inner-city areas (especially industrial ones) were considered as potential redevelopment areas. It is also important to mention the basic outlines of the planning system. This was based on the Planning Act from 1985, which included the necessity of development of the 'general plan' for the municipality (indicating what type of land use was supposed to be allocated in the particular area; it also included the specific types of urban program, which were supposed to be located on specific sites – like e.g. public infrastructure). This system was based on the fact that the only serious investor in the times before 1989 was the Polish state and no major private investments were taking place. Besides these 'general plans', some local municipalities were in possession of 'detailed plans', which indicated almost the entire architectural shape of whole districts. It was possible to plan like this, as – again – only central government was considered as the investor and the plans were not intended to be the planning basis for any private investments.

This 'planning heritage' – when confronted with the development of free market processes – made suburban areas extremely interesting for new development, especially for new housing (Marszał, 1999). But the local municipalities – re-created in 1992 – did very little or nothing to stop this process – in fact, they did not consider this a problem. Still, valid planning documents were considered as a good element of urban policy (which was not a completely wrong way of thinking) but the mayors or city councils did not really think about any type of comprehensive urban management.

The first major change in this field took place in 1994, when the new Planning Act was accepted by the Polish Parliament. It included the idea of abolishing the old plans and an introduction of new types of planning documents. Since this time, the whole area of municipal planning has started to face major changes. The development idea for the whole area of the municipality was supposed to be included in the new type of planning document – the so-called 'study of conditions and directions of urban development of the municipality'. This new type of document – unlike the previous 'general plans' – was not a legally binding document, and as such could not serve as a basis for issuing a building permit. At the same time, the new act introduced 'local plans' – legally binding documents, stating the conditions for urban development in particular sites. But – which was surprising – the new planning act prolonged the validity of old 'general' and 'detailed' plans up to 1999. This allowed the continuation of the previous policy of rapid suburbanization. But what was even more surprising, the municipalities insisted on prolonging these plans for even longer, and – thanks to the further amendments to this act – the 'old' plans in their majority remained valid until 2003! So – for almost fifteen years after the change of the political and economic system – the 'socialist' planning heritage decided about the directions of urban development in Poland! The outcomes of this process seem to be hardly reversible and still the municipalities are trying to follow the urban policy that was developed in the 1980s.

Another change took place in 2003, when another Planning Act was accepted by the Parliament. It did not, however, change the major elements of the system – rather it changed some specific planning procedures and introduced widely the so-called 'administrative decisions', which are intended to serve as legally binding planning documents in the areas where there is no 'local plan' accepted by the local councils (Borsa, Buczek et al., 2003). And – as not many municipalities are interested in the development of these 'local plans' (they are costly in preparation and the procedure of their approval is long-lasting) – at the moment the major part of spatial development in Poland is based on these 'administrative decisions'

(Jaroszyński, Sawicki, 2004). This creates many problems for developers and also many opportunities for corruption. And what is even more chilling, the current planning act – considered as completely imperfect, but introduced just two years ago – is the subject of complete change into the new one. Such a tendency creates great uncertainty for local governments, developers, investors etc., and does not allow for the creation of ordered urban structures.

In effect, many new phenomena have occurred in the spatial situation of Polish cities. Many of them can serve as bad examples of urbanization processes but one can regard some of them as positive ones as well.

Among the bad effects of the changes in the planning situation, one should mention the lack of a legally binding planning document for the entire area of the city. This brings a high degree of uncertainty for many investors, developers or simply land owners, who just do not know what can be done with their land. Such a situation can be found in many of the cities, which – in many cases – blocks the development process. This was also associated with the lack of 'local plans'. As a result, e.g. in the case of the city of Krakow at the beginning of 2004, only 5% of the city area had a legally binding planning document. All the rest of it was ruled – or rather one should say – not ruled – by 'administrative decisions'. This of course created chaos in the building industry and many developers had to wait or hope for a quick decision by the city to issue an 'administrative decision'.

In contrast, some good effects of these changes were also noted. Among them one should mention the processes of re-planning of the old industrial sites, which started the processes of urban regeneration. As mentioned in the following chapters, this process is not really well developed yet, but one can note a change in thinking. This has happened due to the fact that municipal planners have started to see the local problems and not only the general ones common to entire cities. Another good effect of this situation is the appearance of the new planning quality – in these rare cases where the municipalities prepare the 'local plans', they usually include the detailed solutions, made in the regulatory way. This brings the new planning philosophy, and gives hope for the more ordered space in future (Kochanowski, 2002).

Changes in Administrative Structure

As described in the above chapter, the Polish legal planning system has been evolving quickly during the last two decades. Also major changes have taken place in the field of spatial administration. Before 1989, Poland was divided into 49 provinces, and each of them was managed by a government official – the so-called 'voivode' or – simply speaking – governor. Each of these was divided into municipalities of different sizes, which also included cities.

The first major change took place in 1992, when local government reform was introduced. It included the re-creation of local governments and gave them spatial planning power. But – as indicated in the above chapter – they did not make much use of this, except in some of the big cities. Still many competencies – like issuing building permits – were in the hands of the post-socialist structure of 'regional offices', which were the remnants of the voivodeships (abolished in 1975).

Another reform took place in 1998, when the institution of voivodeships was re-introduced. They incorporated the powers of the above-mentioned 'regional offices', as well as the responsibility for managing the public infrastructure (like roads, high schools, hospitals) at the supra-local level. But the voivodeships do not have any spatial planning power, except administrative duties in the field of the building process. The only improvement in spatial development area was noted in big cities – where the mayor became also the head of the gmina. This gave him more power in controlling the building process.

Within the framework of this reform, also the regional administration was reformed. The number of voivodeships was reduced from 49 to 16, and the responsibilities of the voivodes were – to a large extent – limited. At the same time, the regional government was created (it had never existed in Poland before), which became responsible for – among other things – regional planning and development. In its hands there is the duty to prepare the regional development plans and strategies, as well as the preparation of the plans for the metropolitan areas (Buczek, Borsa et al., 2003). But these planning competencies are not very strong, as still most of the power is in the hands of the municipalities. And they are often not interested in cooperation with their neighboring partners. The best example of this is Gdansk and Gdynia – the two cities are not able to cooperate in the field of infrastructure development and also they do not tend to work together in the planning area (Kołodziejski, Parteka, 2001).

Lack of State Intervention in Urbanization Processes

Besides the administrative and legal changes, along with the political and economic transition, the Polish state decided to withdraw from most forms of governmental support for urban development. It was supposed to be replaced by local programs and policies, prepared and implemented by municipalities. The post-socialist governments believed that the best decisions concerning local communities can be made only at the local level. This belief was followed by intended decentralization. But this process related only to duties – not to money distribution. As a result, many of the tasks were transferred to the local and regional governments, but the state did not provide the financial basis for carrying them out. And again, the bigger cities – with a much broader economic basis and more opportunities to attract investors and capital – had a better position in this situation than small urban and rural communities. But even in the case of big cities – as many tasks had to be financed – it was not possible to find enough money even for the necessary infrastructure improvements. In fact, until now the municipalities are struggling with the heritage of the socialist model of urban governance. And – among the main problems – one can find e.g. the degradation of a large amount of the housing stock (never renovated after the war), underdevelopment of infrastructure (especially in the field of water supply and sewage collection, but the main problem is associated with the underdeveloped road network and the very poor quality of many roads) and many others.

As a result of these two processes – decentralization of power and struggling with the socialist heritage – neither of the possible agents (which means the central government and municipalities) had the interest and money to introduce any kind of support system for urbanization (or re-urbanization) processes. Therefore, they were left for the market as some liberal politicians (and also planners) believed that the free market would do its job and clear the situation. But they forgot that the free market looks for the cheapest and most efficient ways of fulfilling economic tasks and does not necessarily take into account the social and environmental aspects of the development process.

As a result, for the whole period since 1989, there has been no single state program aiming at influencing the urbanization processes. The best example is the so-called Urban Regeneration Act – a parliamentary initiative to support the urban regeneration processes. Work on these documents has been conducted for over ten years now, and still no-one is able to predict when it will be finished. At the same time, local municipalities do not intend to influence the urbanization processes, which in many cases have a supra-local dimension. They are usually focused on providing the legal framework for developers as new developments mean new tax inflows. All other planning principles are usually left aside.

Trends in Urbanization

As described in the previous chapter, the current urbanization practices are associated with the domination – or even over-domination – of the free market. At the same time, new products are required on this market –in terms of both housing and other types of urban program. But only in a few cases are they being developed in inner-city locations. All these aspects describe the picture of the urbanization model we have to deal with in Polish cities. This chapter includes a few remarks concerning this model, which – of course – may evolve in a somewhat different way in each of the Polish cities.

Lack of Cooperation between Public and Private Sectors in Urbanization Processes

As stated in the paragraph above, due to the policy of the Polish state and most of the local municipalities, the private sector activities are not influenced in any way by the public sector. This means they are not influenced by financial participation that can change the program or character of the development but of course each project needs to follow the planning regulations concerning the size of the building or the complex of new dwellings, the type of land use and other building regulations.

This situation is associated with the fact that the public sector – again, thanks to the money shortage – is not able to play an active role in the real estate market. Unlike many Western European municipalities, only very few local governments in Poland are able to develop and maintain a large amount of public housing stock. And if they are able to do so, usually they focus on a hundred per cent municipal housing, constructed to house the poorest and homeless families. This means that a great majority of new housing and nearly all other developments are being built with private money only. And of course the buyers have to pay the market price for this. In fact, Polish municipalities used to have a lot of public housing stock but during the last decade most of it was privatised or is under privatization. This happens because the buildings are usually in bad condition and need major improvements. And the municipalities have no money for this, so they usually transfer it for almost nothing into the hands of current users. This shifts the maintenance and up-keep problem to the building communities.

At the same time, all public infrastructure – like schools, roads and so on – is being built with public money only. There is very little experience with public-private partnerships, and only in a few cases do private developers of large-scale projects contribute some money for infrastructure improvement. But this happens only when a new road connection has to be developed in order to connect the site of the project to the existing road network. This happens due to a lack of expertise and experience with cooperation – which is characteristic of both partners. Municipalities are afraid to get involved in public-private enterprises and have no knowledge how to do it safely and properly. At the same time, the private sector is – in many cases – very short-sighted and not interested in the development of the public infrastructure. Such projects need partnerships not only when under construction but also during day-to-day operation. The best example of this tendency is a little housing complex outside Gdansk, where the developer designed a bit of social infrastructure for the inhabitants, including a little basin with a fountain and with a system of lighting-up the buildings. In a short time, it was decided to remove all these elements as neither the municipality nor the housing associations (the formal owners of each of the buildings) were willing to pay for their maintenance.

Housing Needs and Market Responses

Poland is one of the few European countries with still huge housing needs. This relates to both the standard of existing housing and to the number of flats available per capita. This means that both the quality and quantity of apartments need to be greatly increased. Such a situation is the outcome of the insufficient amount of housing constructed in the post-war times and the lack of major modernization efforts in the pre-war housing stock. Therefore, there is a market for all kinds of housing but still the price per square meter of new apartment plays the key role. And – unlike Western European countries – there is no custom of renting apartments. This is costly for the renters and Polish law creates some problems for flat owners willing to rent them out. As a result, only in big cities and in academic centers can one find a larger offer of apartments for rent. But most people – who are looking for housing adequate to their needs – are usually trying to buy a flat on their own. And – apart from some tax incentives (which – by the way – are no longer available) – there is no state or municipal support. The developers have to pay the market price for the land and for all materials and services. As a result, the potential customers have to pay the full price for the newly purchased apartment.

As the price is still the key issue, the developers are trying to offer an adequate product to the customers. The main criterion is the price. In order to make the offer attractive, they try to cut down costs. The only way they can do this is to buy cheaper land. This means that the majority of cheaper housing projects are located far away in the suburbs of the cities, and – in many cases – in the areas of the surrounding municipalities. What is characteristic, these are usually multi-storey houses, not single family houses. Unfortunately, the clients do not take into account other costs of living in the suburbs – like transportation to the centers (there is usually no public transportation), very low quality of roads, lack of social infrastructure and no public facilities, like schools. All of these are in the inner cities or large housing districts developed in the 1970s and 1980s.

At the same time there is some market for the upscale apartments, which is limited to the big cities or to the popular summer resorts. For example, in Sopot or Mikolajki – very popular summer destinations – many of these apartments are purchased by the rich as second or summer apartments. But the regular inhabitants of these cities are very rarely the customers of the developers building these projects. What is characteristic of the city-center projects is that they are usually developed on empty plots and do not utilise the degraded areas – like post-industrial ones. They are also costly and the potential developer has to pay – among other things – also for the clean-up of the area, which makes the project problematic from the financial point of view.

The high cost of land in the inner cities makes building new city-center housing very costly and available only for the rich. Therefore, anyone who would like to buy a modest flat there has to look for a second-hand apartment. This means that most people interested in buying a new flat have to go to the suburbs or take the risk of costly renovation and decide on a second-hand offer. The same relates to single-family housing (Marszał 1999).

New Challenges – Emerging Types of Urban Program

Besides housing, there is a whole variety of new types of urban program that have appeared in Polish cities after 1989. The most important parts of these are new shopping facilities. Apart from them, one should mention offices as well as new leisure and sports facilities.

As far as shopping is considered, it is necessary to remember about the on-going differentiation of clients and market offer. In the mid-1990s the customers were looking for cheap offers, which were provided by large-scale supermarkets, located usually outside the cities. It was necessary to have a car to go there. They were offering cheap products, although at the beginning they seemed to be attractive also for the more demanding clients. In time, this offer was also differentiated, and the cheapest segment of the market appeared. The offer for richer people started to move to inner areas. And – at the end of the 1990s – the new shopping galleries started to appear in the city centers and in their close vicinity. As they gathered the offer for the middle class, their owners were able to meet the costs of running their businesses in the city centers. As a result, the offer on the market is now very diverse, and each market segment tends to locate in a different part of the urban area. Again, this happens with little influence from the municipality – it usually reacts only when a political problem with the local shop-keepers arises.

The same situation can be observed in the case of offices and leisure and sports facilities. New office centers of different standards arise and the most expensive ones are located in the hearts of the cities, while the lower-class facilities are on the outskirts. At the same time, a new (in Polish conditions) phenomenon has arisen – which is the connection between the development zones and transportation nodes. New office complexes arise in close proximity to airports or major transit roads.

Unlike offices, new leisure facilities tend to fill empty spaces within the existing city centers. The attempt to locate large portions of such a program outside city structures was not successful. Both new hotels and multi-screen cinemas are being constructed in the already urbanised areas, usually in association with existing centers or other cultural facilities. For example, in the Gdansk agglomeration there were four major new cinema complexes constructed after 1989, and two of these were located in the central areas of the city. One of them was developed – along with a major shopping center – in the heart of a housing district from the 1970s. Only one – the least successful – was developed outside the urban structures. The same relates to hotels – they are located in the central areas or in the countryside, usually utilizing abandoned palaces and noblemen's houses.

Figure 7.1a   Madison shopping centre in downtown Gdansk
Source: Uwe Altrock

Figure 7.1b   New urban structure for the Young City area in Gdansk
Source: Drawing by P. Lorens, D. Zaluski, S. Ledwon

Figure 7.2a   Plan of Warsaw at the beginning of 16th century
Source: Ostrowski, 2001

Figure 7.2b   New skyline of downtown Warsaw
Source: www.e-warsaw.pl (24.01.2005)

Finally, there are not many new sports facilities but they also tend to be located in the existing urban structures. This comes from the close proximity of their clients – the inhabitants of the city. The population of the suburbs is still not sufficiently numerous to provide a strong clientele for such facilities.

Problems with Degrading Inner Urban Areas

As an effect of the urbanization processes and of the de-industrialization of the cities (which started to take place in the mid-1990s), the inner cities started to suffer from some degradation and urban blight. But – which is interesting – this did not mean a lowering of the prices for the land. The market still indicates that inner city areas are the most expensive ones, even allowing for the fact that most of them need a major clean-up process. Also, in some cases, the land titles are not clear or the plots need a major infrastructure upgrade. The best example of this is Granary Island in Gdansk – the most attractive building site in the whole agglomeration and also the most expensive – but nobody will invest there unless there is a new infrastructure connection to the plots.

Beside these post-industrial sites, there is a shrinking number of undeveloped areas, which are the effect of war-time destruction. The best case is the area around the Palace of Science and Culture in Warsaw, which – once a vibrant city district – is now a huge empty site. But this is one of the few sites like this left – all around one can see new office and hotel towers popping up. But this building boom happens only in Warsaw – all other cities are still waiting for their chance.

Emerging Concepts in Urbanization

Although the free market dominates the urbanization process, it also brings rapid changes in supply and demand. People not only want housing – they are starting to look for housing in some kind of style. They are not fooled any more by 'cheap copies of real cities', sometimes offered by developers. They want to live a really urban life. This comes from the great societal changes that have taken place in Poland since 1989.

Among them one should mention the huge differentiation in housing needs. There is a new group of singles, willing to live on their own and not willing to compromise their standard of living. Another group can be described as young couples with no children or older couples whose children are already mature enough to live on their own – all of them are rediscovering the values of city life. In their cases it is not important to have an apartment as such – they are looking for some special offers. Beside these, one can mention the whole variety of different lifestyles that are tending to emerge in society. All of these decide about the rapid growth of alternative urbanization scenarios (Gaczek and Rykiel 1999).

The Growing Importance of Urban Regeneration

As the societal needs become differentiated, so does the market offer – this is the old rule of the free market. In the pure market-led urbanization in Poland it is also true in regard to the new urban program. At the same time there is a large portion of society not willing to change their living standards but to maintain the existing one. They are usually housed in buildings developed before the Second World War, which – in general – were not maintained properly during the post-war times and did not undergo any major renovation. Therefore, in the last decades Polish cities have started to face another problem – the problem of decaying housing structures, which is starting to be a general problem for society and cities. During the last few years many conferences, research papers and seminars have been devoted to an analysis of this problems and a search for the best possible solutions.

In general, this problem has three dimensions. One of them is related to the already bad technical condition of pre-war structures. In these cases, the renovation efforts should keep as a goal the technical upgrade of the material substance and solving some of the over-population problems. The second dimension is associated with post-war housing, which – in the majority – is prefabricated large-scale housing. Usually it is still not suffering major technical problems (but they will appear soon) but is degraded morally and socially, with many social problems. And finally it is necessary to mention the post-industrial, post-harbor, post-railway and post-military areas still waiting for redevelopment, which – as mentioned in the previous chapter – are usually ready for development but suffer from an underdeveloped infrastructure (Markowski 2004).

The first regeneration programs are usually perpetuated by the private sector, which is interested in developing the new market products in the form of lofts and/or 'stylish' offices or restaurants. But these initiatives are supplemented by 'spontaneous' regeneration – like, for example, acquiring old factory halls for the purposes of artists' workshops. Due to such efforts, the faces of such run-down districts as Praga in Warsaw and Mlode Miasto (Young City) in Gdansk have started to be altered (Lorens 2002).

The Development of Gentrification Processes

Apart from regeneration, in some parts of the cities considered to be extremely attractive, one can observe the starting of the gentrification process. This is possible due to the fact that – after the war – these areas were usually populated by a mix of inhabitants, with both relatively high and low incomes. After the economic and political transformation, when the housing market was re-established (before 1989 selling or buying a flat on the free market was rather complicated, as the majority of the inhabitants did not have a title to the flats – they were considered as 'renters'), some of the city districts started to be considered more attractive than others. This natural process led to the differentiation of prices on the second-hand market of flats. A flat – due to its good location – could cost even twice as much as the same kind and quality of apartment, located in a district considered as unsafe. And – thanks to the same location factor – some of the flats seem to be even unsellable.

But thanks to the introduction of these market mechanisms and the privatization of flats (both municipal and those belonging to the so-called housing cooperatives), it is possible to map the prices of flats and find out which of the cities or districts are more attractive than others. In the case of the most attractive – and most expensive – the prices started to go so high that many of the inhabitants started to sell their apartments in order to buy a more convenient one at a cheaper price somewhere else. This process is not yet very visible as most of the transactions in the market are made due to the fact that the previous owner of the flat had died or decided to move to another city. The results, however, are astonishing – the 'good' districts and cities are getting even better and the 'bad' ones much worse. At the same time, some of these 'bad' districts – due to the attractive location or pre-war architecture of the buildings – are starting to be considered as a potential site of regeneration processes. But – as mentioned in the previous paragraph – these processes are still not well developed.

Conclusions

At the beginning of the 1990s, the World Bank published a report on the Polish cities. Its title said everything: 'Cities without land markets'. At that time such a point was totally justifiable. After fifteen years of economic transformation, however, and after development of the housing and land markets and petrification of the planning conditions, one cannot find it true any more. Now in Polish cities we have land markets but the only rule governing the market is the meaning of location – just like in many other countries. And of course the land owners try to introduce the 'highest and best use' in the plot they possess. But the problem is that this initiative is not necessarily good from the point of view of the entire city or even metropolitan region, where the new developments get dispersed while large portions of land are left undeveloped in the hearts of the cities. If this is not changed, we can soon end up with 'bagel-like' cities – with a ring of urban developments in the suburbs and almost nothing left inside except the historic cities and some shopping and leisure facilities. The worst is that the municipalities do not seem to see this situation as a problem – many of them believe that city centers will defend themselves. The problem is that the free market alone cannot necessarily provide such a solution in a predictable time frame (Lorens 2002).

References

Borsa, M., Buczek, G., Jaroszyński, K., Korzeń, J., Lasocki, M., Świetlik, M. and Szelińska, E. (2003), *Ustawa o planowaniu i zagospodarowaniu przestrzennym.* Warszawa: Urbanista.

Cieślak, E. and Biernat, E. (1975), *Dzieje Gdańska.* Gdańsk: Wydawnictwo Morskie.

Gaczek, W.M. and Rykiel, Z. (1999), 'Nowe lokalizacje mieszkaniowe w przestrzeni miasta.' In: Marszał, T. (ed.). *Budownictwo mieszkaniowe w latach 90. – zróżnicowanie przestrzenne i kierunki restrukturyzacji,* Warszawa: KPZK PAN.

Jaroszyński, K. and Sawicki, M. (2004), *Decyzje o warunkach zabudowy i zagospodarowania terenu.* Warszawa: Urbanista.

Kochanowski, M. (ed.) (2002), *Przestrzeń publiczna miasta postindustrialnego.* Gdańsk: Wydawnictwo Politechniki Gdańskiej.

Kołodziejski, J. and Parteka, T. (eds) (2001), *Ład polskiej przestrzeni. Studium przypadku – Metropolia Trójmiejska.* Warszawa: KPZK PAN.

Kucharska-Stasiak, E. (1999), *Nieruchomość a rynek.* Warszawa: Wydawnictwo Naukowe PWN.

Lorens, P. (ed.) (2002), *System zarządzania przestrzenią miasta.* Gdańsk: Politechnika Gdańska Wydział Architektury.

Lorens, P. (ed.) (2002), *Large-scale urban developments.* Gdańsk: Wydawnictwo Politechniki Gdańskiej.

Markowski, T. (ed.) (2004), *Wielkoskalowe projekty inwestycyjne jako czynnik podnoszenia konkurencyjności polskiej przestrzeni.* Warszawa: KPZK PAN.

Marszał, T. (1999), 'Zróżnicowanie I kierunki rozwoju budownictwa mieszkaniowego w Polsce.' In: Marszał, T. (ed.), *Budownictwo mieszkaniowe w latach 90. – zróżnicowanie przestrzenne i kierunki restrukturyzacji.* Warszawa: KPZK PAN.

Ostrwoski, W. (2001), *Wprowadzenie do historii budowy miast.* Warszawa: Oficyna Wydawnicza Politechniki Warszawskiej.

Sołtysik, M. (1993), *Gdynia miasto dwudziestolecia międzywojennego.* Warszawa: Wydawnictwo Naukowe PWN.

Węcławowicz, G. (2003), *Geografia społeczna miast.* Warszawa: Wydawnictwo Naukowe PWN.

Wojciechowski, E. (1997), *Samorząd terytorialny w warunkach gospodarki rynkowej.* Warszawa: Wydawnictwo Naukowe PWN.

Chapter 8

Urban Development, Policy and Planning in the Czech Republic and Prague

Luděk Sýkora

Introduction

The post-1989 urban change in the Czech Republic has been conditioned by government-led reforms aimed at the establishment of a capitalist system based in a pluralist democracy and a market economy and the integration into international political and economic systems. The establishment of market principles of resource allocation and the growing exposure to an international economy created conditions for the development of spontaneous market-led transformations of economic, social and cultural environment. Urban change has been especially influenced by internationalisation and globalisation, economic restructuring in terms of deindustrialisation and the growth of producer services, and increasing social differentiation (Sýkora 1999). Last but not least the urban development has been impacted by the approach of national and local governments that favoured unrestricted market development.

The political and economic transformation significantly influenced settlement systems and in particular the spatial restructuring within urban areas. The most dramatic changes occurred in the Capital City of Prague and its metropolitan region, where most of the new investments are concentrated. The general metropolitan growth however has been accompanied by internal differentiation within Prague contrasting booming areas with declining areas. Other cities did not have such a favourable position as capital city to attract new investments. However, a number of them succeeded in developing industrial zones and gaining new production capacities offering jobs for local population. Still there are cities whose development is primarily associated with deindustrialisation and decline.

The city governments, with the support of national programmes, attempted to influence the development of cities and their position within the national system. The internal urban transformations have often been left to the operation of free market bound within the framework of traditional physical planning instruments. However, after a decade of transition, many urban governments learned new techniques of urban management and governance and started to apply more sophisticated tools, such as strategic planning. The application of EU programming documents in the

Czech Republic further helped to consolidate urban government measures towards more coordinated and complex solution of urban problems.

This chapter begins with an overview of the development of cities and changes in their urban spatial structure in the Czech Republic after 1989. Special attention is given to the identification of the most pressing urban problems. The later sections are devoted to urban policies and planning at the city level and to the discussion of national government policies and programmes that influence urban change. The final part provides an example of urban planning and policy in the capital city of Prague.

Postcommunist Urban Development: Growth and Decline of Cities and Change in Urban Spatial Structure

In 2001 10.3 million in the Czech Republic inhabited an area of 78,864 km^2. Over 70 per cent of the population is urban and 63.6 per cent of the inhabitants live in towns and cities with a population of over five thousands. The demographic change since 1989 has been characterised by the decline in the total population and an ageing population caused by very low fertility and by shifts in the structure of households with a growing share of single member households and a declining share of couples with children. These changes have been especially pronounced in major cities.

Urban change is associated with the geographic redistribution of population. While major cities loose population through migration, small municipalities gain it. A large part of out-migration is towards suburban areas, especially around Prague and Brno (Čermák 2004). There is a remarkable regional differentiation in housing construction with booming suburban areas, namely around the capital city of Prague, where the wealthiest Czech population is now building new homes. In the districts surrounding Prague, the intensity of housing construction (no. of completed flats per 1000 inhabitants) is nearly three times higher than the national average. However, the transformation in settlement pattern has been rather conditioned by economic change in comparison to demographic change.

There has been a remarkable difference in the dynamics of urban development and urban restructuring between major Czech cities and their regions. The urban growth and decline has been influenced by economic restructuring on the national level and strongly conditioned by the position within the international economy. The variability was especially influenced by the position of individual cities in the hierarchical divisions of labour within the Czech economy being integrated into the international economic system. The potential of cities was given by their inherited economic base, geographic position and attractiveness for new investments. The urban economic restructuring has been characterised by deindustrialisation and tertiarisation and strongly affected by local urban labour markets. While employment in manufacturing and construction declined, the number of employees in services increased. Despite the universal decline in manufacturing, there are still major differences between cities with Prague having less than 15 per cent of jobs in manufacturing while the 3rd largest city Ostrava has 37 per cent. In Prague, and to a certain extent in Brno and some other towns, the decline in manufacturing was balanced by the increase in the service sector. There are, however, also towns and cities that have been severely hit by the economic decline with very limited options for alternative growth.

The capital city of Prague has strengthened its position as a prime national centre and has assumed the role of a gateway, linking the national with international economy (Drbohlav and Sýkora 1997, Dostál and Hampl 2002). The inflow of foreign direct investment and the growth in advanced services confirmed Prague as the country command and control centre. The city is also a major national logistic hub with a huge pool of relatively wealthy consumers. The growth in advanced producer services greatly influenced the structure of jobs, as well as salary levels, and the booming property development, which makes the capital city quite different from the rest of country. The capital city of Prague is the only city where a sufficient number of new jobs were generated to replace the losses from deindustrialisation. There are even structural shortages of labour and low paid jobs, and in a number of instances these jobs are taken by labour migrants from Eastern Europe.

In the Czech Republic, there is no other city that would assume the role of gateway between the international and the local economy. This affects especially the second largest city Brno, where employment in traditional manufacturing quickly declined. Brno aspired to play a more important role than merely being a manufacturing centre. The city, for instance, initiated the establishment of a Czech Technology Park and intended to develop a huge development project of so-called South Centre. Masaryk University in Brno accepts the highest number of new students from all Czech universities. However, in reality the major growth in Brno has been in retail, i.e. the sector that offers only lower level salaries. The city government finally started to attract production capacities to the newly established industrial zone and the city also succeeded to develop as an important logistic hub.

New labour opportunities in other cities were associated mainly with the growth of individual entrepreneurhip, growth in retail sector and state administration. This however, has not been sufficient to cover the decline in industrial jobs. Therefore, all cities, except Prague attempted to attract new foreign investments to supply jobs in manufacturing. In some other cities, there has been strong reindustrialisation. For instance Plzeň has been quite successful with its early offer of land in industrial zone Borská Pole to foreign investors. Consequently the establishment of new production capacities supplied new jobs that were substituting for decline of employment in traditional manufacturing production. Similarly Kladno, a traditional mining and metallurgy centre and the largest town in Central Bohemia, was strongly affected by the decline of its metallurgy base. However, it succeeded in attracting new employers to newly established greenfield industrial zone and also benefited from its proximity to the capital city of Prague and its booming labour market. As these cities could not compete for service jobs they attempted to attract foreign direct investments (FDI) into manufacturing by offering cheap land equipped with necessary technical and transport infrastructure for construction of enterprises, and a cheap and skilled labour force. Despite increasing overall unemployment, the rates in these cities and towns are below national average.

Some cities have not succeeded in the competition for new investments and now exhibit decline and unemployment. Their situation is usually a combination of severe decline of industries inherited from Communism and a low current desirability for new investors due to the bad quality of the physical and social environment, and geographic distance from the western frontier (in the case of Ostrava this is further strengthened by the non-existing highway connection to North Moravia). Cities and towns in old industrial regions in North Bohemia and North Moravia formerly associated with mining, metallurgy and chemical production are those that have been most severely hit by de-industrialisation and have not succeeded to attract new major investments. Their current situation is shaped by economic problems that produce high unemployment. In some cities, such as Havířov, the unemployment is reaching levels over 20 per cent. The economic decline in these cities is not only the question of cities itself but whole regions with a high concentration of heavy industries. Therefore, the base of many of their problems is in the nature of regional economy and has to be tackled by coordinated regional policies and FDI support programs to strengthen their competitiveness for new investments.

The support for economic growth in these areas remains an important task for national economic and regional policy.

Each city and each local labour market has been impacted by a combination of several forces including inherited economic structure, contemporary attractiveness for foreign investors and activity of local governments in attracting them. While all cities have been affected by deindustrialisation, only some benefited from the new developments. In general, Prague quickly adapted as the centre of advanced services, some other cities benefited from reindustrialisation and growth in consumer services. However, there are also cites that were exposed to the severe consequences of deindustrialisation that have not been balanced by growth in other sectors of the local economy. The differentiated external conditions have been decisive for urban development in particular cities.

Major urban changes occurred within the internal space of cities. On the supply side the urban restructuring has been conditioned by the government directed reforms, especially privatisation and price and rent deregulation, which have created conditions for the establishment of urban property markets (Reiner and Strong 1995, Strong et al. 1996). The demand side has been largely differentiated between cities. In Prague, the newly emerged actors in private sector, mainly foreign firms, fuelled the operation of land markets and started to reorganise land use and reshape the historically developed urban structure. This has also happened in other towns and cities, but these developments have been smaller in the extent of changes and have taken other forms. For instance, new office buildings of international standard have been developed nearly exclusively in Prague, while shopping centres have mushroomed over the whole country.

Czech cities are characterised by small urban cores of medieval origin, large inner cities originating with the industrial revolution of the second half of the 19th century, further developing through the first half of the 20th century, and vast areas of new industrial and residential estates from Communist times. The urban growth after 1989 concentrated in the most attractive locations of the city centre, some adjacent nodes and zones in inner city, and in numerous suburban locations. The main transformations in the spatial pattern of former communist cities and their metropolitan areas included (1) the reinvention, commercialisation and expansion of city centres, (2) the dynamic revitalisation of some areas within the overall stagnation in inner cities, and (3) the radical transformation of outer cities and urban hinterland through commercial and residential suburbanisation (Sýkora 1999a, Sýkora et al. 2000). The city centres and suburban areas have been territories with the most radical urban change. Most of the 1990s were characterised by huge investment inflow to city centres causing their commercialisation and decline in residential function, albeit substantial physical upgrading (see for instance Ptáček et al. 2003 for the study of Olomouc). Since the late 1990s, decentralisation occurred with investments flowing to both out-of-centre and suburban locations. Central and inner city urban restructuring involved the replacement of existing activities with new and economically more efficient uses and took the form of commercialisation, gentrification, construction of new condominiums, brownfield regeneration, the establishment of new secondary commercial centres and out-of-centre office clusters. Sýkora (2005a) provides an account of gentrification, Temelová (2004) offers a study of transformation of a former industrial site into new commercial node in Prague-Smíchov, Temelová and Hrychová (2004) discuss how the inner city revitalisation is mirrored in the socially differentiated use of public space. Since the late 1990s, suburbanisation has become the most dynamic process changing the landscapes of metropolitan regions. It brings about a complete reformulation of metropolitan morphology, land use patterns and socio-spatial structure (Sýkora and Ouředníček 2005).

Emerging Urban Problems

Post-communist transformations brought uneven spatial development within cities, redifferentiation of land use patterns and an increase in socio-spatial segregation (Sýkora 1999b) thus changing the formerly rather homogeneous space of socialist cities. The uneven character of post-1989 urban restructuring was caused not only by decline of some urban zones and areas, but also by the investment flowing only to some parts of the built environment, while many areas were omitted. Both decline and growth are causing a number of urban problems. While urban problems are usually associated with economic, social and physical decline, the implications of the uneven character of growth are often omitted. Growth can have negative consequences and contribute to decline in the same geographic area. For instance commercialisation of the city core has negative implications in the sense of population decline, growth of individual car traffic and damage to historical heritage. On the other hand side, growth in one location also has implication for decline in other geographic areas. For instance, booming suburbanisation contributes to the decline in inner city and housing estates.

Figure 8.1a   Gentrified neighbourhood Praha Vinohrady
Source: L. Sýkora

Figure 8.1b   Regenerated offices in Prague-Karlín brownfield area
Source: I. Sýkorová

Figure 8.1c   New condominiums in Praha Černý Most
Source: L. Sýkora

Figure 8.1d   Master planned suburban area in Hostivice near Prague
Source: L. Sýkora

Since the beginning of the 1990s, the central parts of cities have been experiencing the strong pressure of new investments. While these investments contributed to physical upgrading and brought more economically efficient land use, they also contributed to the densification in central city morphology. The higher density and intensity of use contributed mainly to increased use of the central parts of cities including rapid growth in car traffic and consequent congestion (especially critical has been the situation in Prague). The disappearance of green spaces in inner yards is another effect of this process. Furthermore, as Czech cities have medieval cores there were numerous conflicts between investors and the protection of historic buildings and urban landscapes. Commercialisation, i.e. the increase in the share of commercially used floorspace led to the rapid decline of residential land use in inner cities and the out-migration of residents. Consequently, there are now blocks of central city properties without any residential function – a problem known from western cities.

There are two particular zones within Czech cities that are currently threatened by downgrading. These are old industrial districts and post Second World War housing estates. Inner urban industrial areas are affected by economic restructuring and are becoming obsolete. Old buildings, contaminated land, and complex ownership patterns complicate the regeneration of these areas. Furthermore, in many cities and locations there is virtually no interest in their redevelopment. Brownfields left by deindustrialisation, and in some cities such as Olomouc by demilitarisation, are becoming one of the major problems areas for many Czech towns and cities. Up to now there have been rather scarce examples of the reuse of former industrial areas, namely associated with the redevelopment driven by commercial functions in locations near city centres, such as Smíchov in Prague (Temelová 2004), or specific functions, such as the construction of new multipurpose sport and cultural hall Sazka Arena in Prague Vysočany associated with the World Hockey Championship 2004.

Another problem area are housing estates of large multifamily houses constructed with the use of prefabricated technology during the 1960s–1980s for tens of thousands of inhabitants. Their life span and technical conditions call for regeneration; otherwise this will lead to physical and social decline. Due to the extent of housing estates and current out-migration of more wealthy people from them, their areas may present one of the largest concentrations of physical and social problems in coming decades. This may concern in particular those cities whose labour markets are strongly affected by economic decline. The population affected by unemployment usually concentrates in housing estates. Rent arrears and limited financial resources of the owners contribute to low level of maintenance, disrepair and physical dilapidation. Even in booming cities, there is an ongoing remarkable differentiation between housing estates. The residential areas that are well located on public transportation and near green areas are perceived as good living addresses and attract new investments into apartment houses, offices and retail facilities. However there are also residential districts with a higher concentration of manual workers and with worse accessibility by public transport, and they show significant signs of decline.

The major growth in post-communist metropolitan areas is concentrated in the suburban zone. The future of brownfields, housing estates and suburbs is interlinked together. If brownfields and housing estates are omitted and get on the spiral of ongoing decline, firms and wealthier people are more likely to leave for suburbs, while inner cities will be characterised by dilapidation and decline.

Suburbanisation itself can become a major problem. The compact character of the former socialist city is being changed through rapid commercial and residential suburbanisation that takes the form of unregulated sprawl. New construction of suburban residential districts is fragmented into numerous locations in metropolitan areas around central cities. Non-contiguous, leap-frog suburban sprawl has more negative economic, social and environmental consequences than more concentrated forms of suburbanisation. The societal costs of sprawl are well-known from North America and Western Europe and now threaten sustainable metropolitan development in the Czech Republic. This concerns not only residences but also new commercial facilities. For instance, suburbanisation of retail facilities has completely reshaped the pattern of commuting for shopping. While in 1990s, most retail was concentrated in central city shopping areas and in secondary centres within cities, at present a large share of shopping is realised in suburban hypermarkets and shopping malls, where people travel by car from the inner city. A very specific example is the city of Brno, where most new shopping facilities were built south of town while most of new suburban residential districts are in naturally valuable areas north of town. Consequently, people commute to shop through the inner city contributing to traffic congestion. Another major impact of suburbanisation is in the field of spatial mismatch in the distribution of jobs in metropolitan areas. Suburban jobs are namely in retail, warehousing and distribution with low paid employees taken by people from the inner city and surrounding region. On the other hand suburban areas are now becoming home of wealthy population that commute to their office jobs in central and inner cities. Therefore, there is developing spatial mismatch between the location of jobs and residences, contributing to increased travel in metropolitan areas and consequent effects on the quality of environment and life. The outcomes of rapidly developing suburbanisation in terms of spatial distribution of people and their activities in metropolitan areas form conditions that will influence the life of society for several generations. Therefore, patterns of urbanisation in metropolitan areas shall become important targets of urban and metropolitan planning and policies that intend to keep a more compact urban form.

Figure 8.2a   Sprawling houses in suburban areas of Praha in Central Bohemian region

Source: L. Sýkora

Figure 8.2b   New out-of-town shopping complex Olympia in Plzeň (Pilsen)

Source: L. Sýkora

Figure 8.2c Průhonice – Čestlice retail and wholesale zone
Source: Geodis Brno

Figure 8.2d Rudná logistic park
Source: Geodis Brno

The post-communist cities are also being impacted by increasing segregation. With growing income inequalities and established housing property markets, local housing markets are divided into segments that are expressed spatially (Sýkora 1999). Wealthy households usually concentrate in city centres, high status inner city neighbourhoods (both apartment housing and villa neighbourhoods and garden towns) and increasingly move to new clusters of inner city condominiums and especially to newly built districts of suburban housing. Less wealthy households concentrate in inner city zones of dilapidation usually associated with declining industries and brownfield formation, and in some post Second World War housing estates especially those originally built and allocated as enterprise housing where larger share of blue collar workers concentrate. A specific urban social problem is the segregation of parts of the Roma population in some cities, where they are intentionally allocated to local government housing in poor condition. Some local government purposefully built shelters for municipal tenants that do not pay rent and move them into this type of very simple housing that is usually segregated on the edge of urban areas. The processes of the separation of the wealthy, and the segregation of poor populations contribute to a changing spatial distribution of population according to social status, growing socio-spatial disparities, and can contribute to the weakening of social cohesion in our cities. The segregation processes are relatively slow; however, once started it will be difficult to later solve its undesirable consequences. Cities with high social disparities and social conflicts are not desirable places to locate new investments and thus social problems can threaten their economic viability and further add to the vicious circle of socio-economic decline.

The post-communist urban development is characterised by an uneven impact on urban space. Most politicians see this as a natural outcome of market mechanisms that are creating economically efficient land use pattern. However, the spatially uneven development can in the future threaten economic efficiency, social cohesion and environmental sustainability. The question of social justice and social cohesion, issues of environmental impacts and sustainability, and more balanced spatial development have been up to now rather subordinated to the preferences given to economic growth. Urban governments could attempt to stimulate investment activity in less preferred locations to distribute the benefits from the growth and development more evenly across the urban territory. In a number of cases, cities need support from the national government to solve some of the most severe problems. The urban problems, however, currently are not among the issues of political and public debate on the national level. Some attention has been given to the decline in post-war housing estates and to the regeneration of brownfields. Most urban problems are, however, seen as local in their nature and left to local solutions.

Urban Development Policies and Planning

In the Czech Republic, the responsibility for urban development rests primarily with city governments. Urban problems are tackled by local governments, which are in some instances supported from national government programs. There is no explicit national urban policy or approach toward cities and their problems. The general conditions for the operation of city governments in the field of urban problem solutions are provided by a general framework for local government system, local government finance and physical planning (Balchin et. al. 1999, Blažek 2002, Maier 1998, Sýkora 1999).

Table 8.1 Structure of municipalities according to population size in 2001

Population size of municipality	No. of municipalities	Population 1.3.2001	Share of country population (%)
-499	3702	868511	8.5
500 – 999	1280	893592	8.7
1000 – 1999	652	903757	8.8
2000 – 4999	363	1118510	10.9
5000 – 9999	130	898301	8.8
10000 – 19999	68	965102	9.4
20000 – 29999	27	678538	6.6
30000 – 49999	14	541501	5.3
50000 – 69999	9	514819	5.0
70000 – 89999	3	237841	2.3
90000 – 109999	6	582307	5.7
110 000+	4	2027281	19.8
Czech Republic	6258	10230060	100.0

Source: Czech Statistical Office, Census 2001

The solution of urban problems including the use of national and supra-national (EU) support is highly dependent on the rights, responsibilities and actual activity of municipal (city) governments. The possibilities of local governments to influence urban and metropolitan development depends not only on the their rights assured by law, but also on their actual strength in terms of financial sources and human capital. Territorial administration reflects the historically formed settlement pattern. The settlement structure of the county is very fragmented with cities surrounded by a large number of small settlements with administratively independent municipal governments. Consequently, the Czech territorial administration is characterised by huge fragmentation. The country consists of 6,258 municipalities (obec) and 14 regions (kraj) both with elected representations. The capital city of Prague and 16 so-called statutory towns can be further subdivided into boroughs. 60 per cent of Czech municipalities have less than 500 inhabitants and further 20 per cent population between 500 and 1,000 (Table 8.1). 90 per cent of municipalities have population below 2000. There are four major cities with population over 150,000 inhabitants: Prague (1169 thousands inhabitants), Brno (376), Ostrava (317) and Plzeň (165). A cluster of six cities with population between 90–105 thousand inhabitants follows: Olomouc (103), Liberec (99), České Budějovice (97), Hradec Králové (97), Ústí nad Labem (95) and Pardubice (91). All these cities are regional capitals. Two remaining regional capitals (Karlovy Vary and Jihlava) are smaller with population slightly over 50 thousands inhabitants.

The inner urban problems in individual cities can be overviewed and tackled (at the local level) by a single municipal government as the urbanised area of cities in the Czech Republic is covered by one local government jurisdiction. Cities are over-bounded, i.e. their administrative territory is larger than the built-up area, and beside the core city also involves a bundle of small village type settlements and agricultural land. A different situation concerns metropolitan development in the functional urban region, i.e. in the area that is tightly linked through the commuting for work, services, education, culture, etc. The area extends far behind the administrative boundary of the core city. Metropolitan areas do not exist as independent administrative units in the Czech Republic. They consist of core cities and a large number of usually small municipalities ranging from villages of few hundred inhabitants to small towns with population around ten thousands. Local governments of core cities can govern the spatial and land use development in the city itself and the small part of suburban zone at their territory. However, they can not influence development behind their administrative boundary. The development there is in the hands of a large number of local governments of usually very small municipalities. For instance, in the hinterland of Prague, there are 171 municipalities and in the hinterland of Brno, 137 municipalities. The fragmented metropolitan decision-making is becoming particularly important with rapidly developing suburbanisation that is in some instances taking the form of sprawl. The coordination of metropolitan development rests on regions, whose priorities have not up to now included issues such as sprawl. Furthermore, the country's largest Prague metropolitan area is under the government of two Prague and Central Bohemian regions, who do not cooperate in the field of common metropolitan development. The metropolitan development is thus now based on the competition between the core city and a large number of suburban municipal governments.

The Czech municipality is an independent legal and economic entity, which takes decisions and bears responsibilities on its own behalf. It has its own means and financial sources, which manages independently according to conditions given by Municipal Act. Municipalities have a right to acquire, dispose and manage municipal property, adopt municipal budget, establish legal entities, adopt a municipal development program, approve local physical plan and issue municipal decrees valid on its own territory. The capital city of Prague and statutory towns can approve a local decree called *Statut*, which divides the municipal territory to districts or quarters, establishes second tier of local government (boroughs) and specifies the decentralization of responsibilities from the municipality (the central city government) to its boroughs. For instance Prague is divided into 57 boroughs.

Municipalities shall ensure municipal development in accordance with interests of their inhabitants. To achieve that they can allocate finances to achieve their goals, and they can use municipal real estate and other property to promote local development and cooperate with other municipalities, state administration and private sector. They are obliged to maintain local streets and roads, care for elementary school facilities and social and health services, maintain water supply, savage disposal and waste management, etc. These services can be provided by municipal enterprises financed through municipal budget (budgetary organizations), by private enterprises established and owned by municipality or in cooperation with private sector firms. One of the characteristic features of service delivery by municipalities during the 1990s was the withdrawal of municipalities from direct service provision and an increasing share of service delivery by private firms.

The basic policy and planning documents are the municipal development program that specifies long-term priorities of socio-economic development, and the medium-term physical plan and the municipal budget that specifies financial and in particular investment allocation in the short-time perspective. Since the beginning of the 1990s, an increasing number of cities have prepared or are currently preparing municipal development programmes that are usually called strategic plans. Strategic plans are the main policy documents of urban governments. The strategic planning is often used in medium-sized and large cities to identify main priorities in economic, social and environmental development through collective bargaining among elected representatives, private entrepreneurs and citizens. It is gaining an increasing importance in the decision-making as a process-oriented strategy based on communication and consensus among stakeholders and the identification of common objectives important for partnership and integration of top-down and bottom-up approaches. There are, however, various forms of strategic planning that differs according to the roles of experts and other participants, methods of problem definition and whether they adopt process or product oriented approach. Maier (2000) recognises two distinct modes among the actual strategic plans: first, the rational, expert-based and product-oriented plans and, second, the visionary, participatory and process oriented planning (Maier 2000). Strategic planning is seen and used as a pro-active type of approach in local governance and stems thus in a contrast with physical planning that is based on the specification of limits for the development (Maier 1998). Strategic planning usually also specifies the ways of using or bidding for national and now EU funds from various policies and investment pockets. For instance, the Strategic Plan of Prague served as a base for the preparation of Regional Operational Programme and later Single Programming Documents for Objectives 2 and 3 to apply for EU Structural Funds (ERDF respectively ESF) support in 2004–2006. Strategic planning helps to integrate and co-ordinate municipal policies and investment priorities in many fields such as physical planning, transport policy, etc.

into one coherent framework, allocate responsibilities for particular fields, and find internal as well as external financial sources to implement approved development priorities. It also has direct implications for the construction of annual municipal budgets linking long-term visions with actual annual allocation of finance and realisation of actual projects.

Physical plans are the major instrument for cities to control the territorial development of their municipality, including the location of new developments, types of constructed building, relations between different function, main infrastructure, etc. Planning documents can have the form of a regional plan (covering the whole metropolitan area), a general land-use plan for a municipal area and local regulation plan for an inner urban zone. The principal authority responsible for procurement of urban physical planning documentation is at the municipal (city government) level and the physical plans are approved by Municipal Assemblies. The procurement of regional physical plans is at the regional level and plans are approved by Regional Assemblies. Regional physical plans can cover and regulate spatial development across many municipalities and coordinate development in metropolitan areas and urban regions. An exception is Prague's metropolitan region that extends across two regional governments: the capital city of Prague has itself a status of region and the surrounding area is under the government of Central Bohemian region. These two regional governments would have to co-operate to achieve an integrated metropolitan strategy, which is not the case at the present. The approved regional plans are binding for the land use plans of municipalities. The approved municipal plans are binding for the elaboration of development projects and the decision-making concerning planning permit.

Urban development is strongly influenced by the financial situation of local governments and thus dependent on the system of local government finances. The application of various national government policies towards cities and urban problems requires co-financing from municipal budgets. Furthermore, there are urban problems, whose solutions are not supported from national sources, and cities attempt to combat them using own financial sources. Therefore, the system of local government finance creates an important framework for the fulfilment of policy goals of urban governments. The Czech system of local government finance has changed several times since the beginning of the 1990s (Blažek 2002, Surazska and Blažek 1996). The current system is based on the sharing of revenues from selected taxes on a per capita principle. The share of municipalities from collected taxes is 20.59 per cent of revenues from the value added tax, business tax and personal income tax. The redistribution from central budget to municipalities is arranged according to the population size of municipalities. Large cities have larger income per capita in comparison with smaller towns and municipalities reflecting their role of centres for surrounding areas. In addition, municipalities directly obtain 30 per cent of taxes derived from personal income tax from entrepreneurs who have permanent residency in a municipality (70 per cent of revenues from this tax are transferred to the central budget from where the share allocated for local governments, i.e. 20.59 per cent is redistributed back to municipalities). This revenue stimulates municipalities to support economic development. Municipalities have negligible revenues from property tax and some local fees. The property tax represents on average only about two per cent of total municipal revenues. Municipalities can, however, differentiate the property tax level in various parts of their territory to influence development in particular areas or zones.

Czech municipal and in particular city governments are important investors. On average, Czech municipalities allocated nearly 40 per cent of their budget into investment during 1993–2001 (Blažek et al. 2003). This is quite a high level of capital investments by local governments in comparison with other countries. During the first years of transition local governments built neglected or even non-existent technical infrastructure. This effort was often financed from sales of municipal real estate and promoted by central government subsidies. Towards the end of the 1990s many municipalities and especially medium and large cities started developing new industrial zones, purchasing land and providing it with infrastructure. Their goal was to offer these areas to both Czech and foreign investors and thus provide jobs to local citizens and encourage local economic development. Czech municipalities are entitled to borrow money and issue communal bonds and they have used this opportunity to cover part of their capital investments. The financing is usually provided by a combination of their own resources (sales of municipal assets), grants from the state support programs, and bank loans that are necessary as the allocation of state grants requires co-financing from municipality and own sources are not sufficient to cover the major investment in time of its realisation. Some municipalities used public private partnerships for the reconstruction or construction of municipal infrastructure and real estate thus avoiding direct bank loans.

National Policies and Programmes with Impact on Urban Development

There is no explicit national urban policy and planning in the Czech Republic and no integrated national government framework or approach toward cities and their problems (Sýkora 2005b). While some problem areas, such as housing estates and brownfields, have been detected in government policies, the overall national urban policy and planning aiming at balanced economic, social and environmental development that would bring benefits to all urban citizens and entrepreneurs and provide chances for all types of cities and individual neighbourhoods is missing. The main national government policies that have effects on urban development are housing policy (Eskinasi 1995, Lahoda 1999, Lux 2003, 2004, Sýkora 1996 2003), regional policy (Blažek 2001, Blažek et. al. 2003), EU policy of economic and social cohesion (Blažek and Vozáb 2004), environmental policy and support for foreign direct investments (Uhlíř 2004).

According to the Conception of Housing Policy, the main housing problems include low levels of housing affordability, spatially uneven distribution of housing stock, and undermaintenance and dilapidation of housing (MMR 2001). The conception declares that the housing needs of inhabitants appear on the local level and therefore the role of local governments in housing should be strengthened. The general availability of housing can be improved by better land policies of local governments. There are limited possibilities to improve the affordability of owner-occupied housing, due to the large disparity between construction costs and households incomes. The central government attempts to increase affordability through the development of new legislation for non-profit rental housing. Another priority is the care for existing housing stock, its modernisation, repair, reconstruction and regeneration of whole housing areas in major cities. The state provides financial support for housing consumption (support to housing saving and to mortgages) as well as for the construction of new housing. There are several housing policy programs that are targeted to municipalities and have a strong impact on urban development especially in larger cities (for details see Sýkora 2003, Lux 2004). The state subsidises production of new municipal rental housing, housing for elderly, and provision of technical infrastructure for all kinds of housing construction. Furthermore, a number of programmes are aimed at reconstruction and modernisation of housing stock to solve problems with dilapidated housing stock and with structural problems of houses built with prefabricated technology. These programmes can help with the regeneration of post-war housing estates.

The national regional policy involves several programmes with an impact on urban areas (MMR 2003). These programmes aim to support the provision of infrastructure for SMEs and tourism and thus combat unemployment. Support is given to municipalities in economically weak and structurally affected areas and in two NUTS 2 regions: Northwest and Moravia-Silesia (Ostravsko). The support is used for brownfield regeneration and the preparation of small industrial zones. In 2004, a new programme was launched for the revitalization of building complexes used until recently by the army (barracks).

The actual development policies applied in individual cities are influenced by the possibility to draw financial support from Objective 1 of the EU Policy of Economic and Social Cohesion. Since the end of the 1990s, the Czech Republic has adopted pre-accession programmes based on principles similar to those of EU cohesion policy (Blažek and Vozáb 2004). Following the accession, the Structural Funds programmes are being implemented via the National Development Plan of the Czech Republic for 2004–2006 and in particular via the Joined Regional Operational Programme for 2004–2006, which addresses the development priorities of seven cohesion regions other than Prague (the Capital City is not eligible for support from Objective 1, however, it can draw funds according to Objective 2 and 3 – see section of planning in Prague). The priorities affecting urban development are regional public transportation and the regeneration and revitalization of the deteriorating urban areas.

Urban development and the resolution of economic problems of declining cities in the Czech Republic are strongly influenced by the support for development of industrial zones. This programme is carried out by the CzechInvest agency established by the Ministry of Industry and Trade for the purpose of facilitating support for the direct inflow of foreign investments into the Czech Republic. The programme forms part of the system of investment incentives for large investors. Within the scope of investment incentives, the government offers large investors tax relief, support for the creation of new jobs and staff training and re-training, and support for the development of industrial zones. The system of investment incentives was originally only intended for manufacturing industry; however, from 2000, eligibility was expanded to include companies in the area of information technology and strategic services (centres for software development, customer support centres, service centres, centres of applied company research, and so forth) with the aim of increasing the competitiveness of the Czech Republic and enhancing the profile of the foreign investments. The crucial part of the system is support for the preparation of large industrial zones for strategic investors. The programme of the development of industrial zones has been in operation since 1998. In 2001, the programme was expanded from greenfield industrial zones to the regeneration of derelict industrial land. The programme provides public financial support to municipalities, regions, and developers in the form of financial subsidies and free-of-charge or low-price transfers of state property for the preparation of industrial zones with an area of more than 10 ha. To date, the programme has stimulated decentralised urban development with only negligible support for brownfield regeneration.

Brownfield regeneration, compact city, and anti-sprawl arguments all play an important part in the State Environmental Policy and the Sustainable Development Strategy for the Czech Republic. However, the implementation of the declared aims to protect suburban land against sprawling tendencies, to stimulate inner city regeneration, and to support the integration of public and private transportation rests on the policies and investment priorities outside the Ministry of Environment itself.

Urban Development and Planning in Prague

Prague is a capital city and with a population 1.2 million also the largest urban centre which dominates the settlement and regional system in the Czech Republic. The city accounts for 12 per cent of the population of the Czech Republic, 15 per cent of jobs and over 20 percent of GDP. Prague is both a municipality and region and the municipal assembly of Prague serves as both a city and regional government. Prague is also one of 8 NUTS II regions in the Czech Republic. Furthermore, the city of Prague is a core of a metropolitan region, which however, has no formal administrative delimitation. Therefore, the planning of Prague's development includes several territorial levels of planning: municipal (the territory of Prague itself), metropolitan (Prague and its metropolitan region), regional/national (Prague as a part of the national regional system) and regional/European (Prague as a part of European regional system).

At the municipal level, two citywide planning documents – the Master Plan and the Strategic Plan – have been under preparation from the beginning of 1990s. The Strategic Plan specifies long-term priorities of socio-economic development and the Master Plan is physical plan that specifies the spatial arrangement and land use in the medium-term perspective. The draft of the Strategic Plan, which specified strategic aims and programs (but did not provide the description of mechanisms of their implementation), was approved by the City Assembly in autumn 1998. In 1999, the Assembly of the Capital City of Prague approved two important documents governing urban development: a program of implementation of (selected) strategic priorities of city development in 2000–2006 (in June 1999) and a new Master Plan (September 1999). The Strategic Plan of Prague was finally approved in May 2000.

The Strategic Plan of the Capital City of Prague formulates the long-term programme of city development for 15–20 years. It declares the will of Prague not to leave the future fate of the metropolis just to spontaneous development. It is intended to purposefully manage the process of urban change and to co-ordinate the decision-making processes of city administration with the numerous interests, activities and resources of various actors in the city. It is a commitment to fulfil the strategic vision of Prague and define paths to prosperity for the city, to a healthy and cultural environment, and to the preservation and development of those values, which make Prague one of the most beautiful cities in the world. The Prague Strategic Plan is not only an overall vision of the city's future, but it is an agreement between politicians, specialists, corporate sector representatives and citizens of Prague about what the city wishes to achieve in the next two decades and how to proceed in the solving of problems in particular spheres. The Prague strategy is based upon the city's strengths, especially its unique character and spiritual, intellectual and cultural traditions, quite exceptional natural and urban quality, its economic and human potential, advantageous position in the heart of Europe, its good reputation, and its attraction for foreign visitors.

The strategic concept for the City of Prague focuses on five main themes, containing a system of mutually inter-linked strategic directions, aims, policies and programmes.

- Successful and respected city (The role of Prague and the city's economy): Prague aims to become a successful, competitive and respected city with a strong and modern economy generating wealth for its citizens, offering prosperity to entrepreneurs, and generating financial resources to public projects.
- Kind and contented city (Quality of life): Prague wishes to be an attractive city of satisfied citizens and visitors. It is determined to provide a good quality of city life in safe and well-balanced communities with equal opportunities.
- Attractive and sustainable city (Quality of the environment): Prague endeavours to achieve a high quality of both natural and urban environment, while observing the principles of sustainable development. It wants to substantially reduce pollution in the city and create a balance between human settlement and landscape, in order to become a clean, healthy and harmonious city.

- Functioning city (Transport and technical infrastructure): Prague aims to modernise, develop and operate its transport and technical infrastructure to support a well functioning city, its economy, ambition and development. Prague's infrastructure should always be modern, reliable, efficient and environmentally friendly.
- Dynamic and welcoming city (Management and administration): Prague wishes to be a city of dynamic and open administration, efficient in providing services and protecting public interests, co-operating with others along the principles of partnership, thus enabling an active role by others and supporting citizen participation in community management and development.

The strategic priorities of the capital city of Prague are:

- to create a favourable entrepreneurial environment;
- to support science and education;
- to protect and develop the city's cultural and urban values;
- to achieve an efficient economy for all resources (nature, land and property, water, energy and finance) with respect to sustainable development;
- to develop a polycentric structure of the city;
- to build the Partnership for Prague between the public sphere, private sector and citizens.

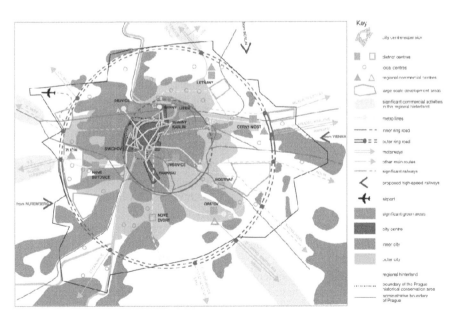

Figure 8.3   Prague: from monocentric to polycentric city structure
Source: Strategic Plan for Prague 2000

The Master Plan of the Capital City of Prague specifies allocation of functions in the city and regulates the development process. The main principles of Prague's spatial development expressed in the Master Plan are:

- Compact city structure (extension of the boundary of compact city so it allows controlled city growth);
- Deconcentration of functions from the city centre through the extension of the city centre and the creation of secondary centres;
- Location of regional shopping centres in outer city;
- Residential areas for medium-rise apartment housing located within the compact city and low-rise single-family housing in outer city locations adjacent to the compact city and existing settlements in outer zone;
- New areas for short-term recreation in outer city;
- Concentration of development to areas with a good accessibility by public transport and in outer city by railway;
- Extension of underground system;
- Construction of inner city ring road and outer city express road.

The regional physical plan for Prague's Metropolitan Region was under preparation in the 1990s as a physical plan of large territorial unit and remained in the draft version (Sýkora 2002, Maier 2002). The plan was intended to guide the long-term development in Prague metropolitan region with respect to the cultural and environmental values of its landscape. It specified the development of main corridors of transport and technical infrastructure so it is not in conflict with residential areas and zones of protected natural environment. The plan also attempted to direct development to (1) selected growth zones along main transport corridors and areas with good transport infrastructure and to (2) zone adjacent to the compact built-up environment of Prague and to larger settlements in the suburban zone which have sufficient social infrastructure (education, health, cultural facilities). The plan aimed to co-ordinate development between the city of Prague and adjacent (mostly small) municipalities. The elaboration of the plan for Prague's Metropolitan Region was under the supervision of the Ministry of Regional Development. With the establishment of regional governments, the question of metropolitan planning now rests on the mutual agreement of two regional governments of Prague and Central Bohemia. The strategic planning document – Development Programme for Central Bohemian Region – is focused only on issues of its own area, without Prague taken into account, and omits spatial development. For instance the issue of suburbanisation in Prague metropolitan area is mentioned only in a minor chapter on tourism and cultural heritage as a potential threat, however, it remains unnoticed in other parts on economy, environment, etc. The two regional governments of Prague and Central Bohemia would have to cooperate to achieve an integrated metropolitan strategy, which is not the case at present.

Since late 1990s, the City of Prague has been involved in the preparation of policy and planning documents that will allow it to bid for EU funds. During 1998–99, the city prepared the Regional Development Strategy (nearly identical with the Strategic Plan) as a part of the national regional development planning and policy and the Regional Operational Program (ROP) as its input into the National Development Plan of the Czech Republic. The ROP contained a description and analysis of the city, SWOT analysis, specification of strategic objectives, description of priorities and programmes, a financial plan, and the outline of implementation institutions and procedures. The main goal of ROP was "to increase the quality of life, to extend and reconstruct technical infrastructure and transport systems and to develop city potential so Prague will become a dynamic metropolis of a future member state of the European Union". Four priorities were selected to fulfil the goal:

- functioning and sustainable urban transport;
- reconstruction and development of technical infrastructure;
- the development of human resources and non-material conditions of competitiveness;
- integration of Prague to European structures.

The Prague's ROP could not be included into the Joined Regional Operational Programme that is part of the National Development Plan of the Czech Republic. Prague as a region significantly exceeds the eligibility criteria for support according to Objective 1 (that is, 75 per cent of EU GDP per capita). Prague has therefore prepared two single programming documents (SPD). The first SPD, prepared for Objective 2 (economic and social conversion) and supported by the European Regional Development Fund (ERDF), concentrates on the development of transport infrastructure and on brownfield revitalization. The second SPD, for Objective 3, concentrates on employment and education and receives support from the European Social Fund (ESF). Both documents were prepared for a shortened programming period of 2004–2006. The main reasoning underlying the application within Objective 2 is that, although Prague is a growing and wealthy region, it suffers from intra-urban disparities. There are large areas affected by industrial decline and support from public sources is needed for their conversion. The territory for which the support is asked via the programming document for Objective 2 covers 40.7 percent of the Prague territory with 31 percent of the population. The main priorities specified in the programming document for Objective 2 are the Revitalization and Development of the Urban Environment (transport systems, technical infrastructure, regeneration of brownfield sites, the generation of labour opportunities, and a better quality of life on the housing estates) and Urban Competitiveness and Prosperity (partnership of public, private, non-profit, and research and development institutions, support to SMEs, development of strategic services to support an information society). This programming document was prepared by the City of Prague in association with the Ministry for Regional Development. The other single programming document for Objective 3 was prepared by the Ministry of Social Affairs in association with the City of Prague. The main goals are employment generation, social integration and adaptability, education, and equal opportunities for both genders.

Conclusions

The first half of the 1990s was characterised by a minimal involvement of governments in urban development (Rehnicer 1997). The decisions of both the central government as well as local politicians were grounded in a neo-liberal approach, which saw a free, unregulated market as the mechanism of allocation of resources that would generate a wealthy, economically efficient and socially just society. Politicians perceived the state as the root of principal harms to society and the economy in particular. The crucial role of the government was to reduce government involvement in as many matters as possible. The urban planning and policy was perceived as contradictory to the market. Short-term, ad hoc decisions were preferred to the creation of basic rules of the game embedded in a long-term plan, strategy or vision of city development. Only towards the end of the 1990s, strategic plans of the city development attempted to formulate more complex views of urban development and city governance. The urban government learned the main principles of urban governance, policy and planning in democratic political system and market economy. The urban planning system was kept in operation and thus helped to regulate smoother development in cities (Sýkora 1995). Basic policy and planning documents, such as strategies of urban development, physical plans, city housing and transport policies and other, were prepared, publicly discussed and approved by local governments during the 1990s. Professionals and politicians learned to pay attention to urban competitiveness, cohesion and sustainability, learning that these issues are high on the European urban agenda. The procedures used in the EU significantly impacted urban planning, policies and programmes including their implementation and evaluation, and urban governments now use benchmarking to monitor and assess the results of their own policies.

There are still weaknesses in contemporary urban policy and planning in Czech cities (Sýkora 2002, 2005b). The first issue concerns the non-existence of a common and coherent national framework that would identify problem areas and attempt to formulate integrated nation-wide cross-sectoral policies and programmes targeting urban questions. There are various sectoral policies with impacts on cities. However, their outcomes are not discussed in any coherent framework. Their organisation and financing is organised along ministerial and sectoral divisions and they sometimes contradict each other. For instance, most of the support to FDI flows into greenfield locations, while the State Environmental Policy declares that the development should be directed towards inner city revitalisation and restrict sprawl outside the compact cities. The co-ordination of various policies and initiatives could more effectively solve existing problems and save some of the allocated public funds.

City governments have high autonomy concerning their own urban planning and policies. After the turbulent transition years, some local governments are realising that a long-term, holistic and complex vision of urban development can be a backbone for city stability and prosperity. In pursuing some of their own agenda in urban development they use support from central government grants and subsidies, namely in the fields of housing and industrial zone development. However, many cities and towns still do not use an adequate marketing/promotion strategy and a land and real estate policy. The sustainability principles are more in the area of rhetoric than implementation. The top-down approach in urban policy and planning still prevails. The cooperation between city governments and private sector often miss a coherent framework. The public awareness has increased and thus urban governments pay more attention to the voices of NGOs and to public participation. However, in many cases confrontation rather than cooperation prevails in communication between the city officials and non-governmental non-profit organisations.

Acknowledgement

The support provided by the Grant Agency of the Czech Republic, project no. 205/03/0337 'New Phase of Regional Development in the Czech Republic' and the Ministry of Education, Youth and Sports of the Czech Republic, project no. MSM0021620831 'Geographic Systems and Risk Processes in the Context of Global Change and European Integration' is greatly acknowledged.

References

Balchin, P., Sýkora, L. and Bull, G. (1999), *Regional Policy and Planning in Europe*. London: Routledge.
Blažek, J. (2001), 'Regional development and regional policy in the Czech Republic: an outline of the EU enlargement impacts.' *Informationen zur Raumentwicklung*, No. 10.
Blažek, J. (2002), 'System of Czech local government financing as a framework for local development: 12 years of trial and error approach.' *Acta Universitatis Carolinae Geographica* XXXVII (2), pp. 157–174.
Blažek, J., Přikryl, J. and Nejdl, T. (2003), 'Capital Investment Funding in the Czech Republic.' In: Davey, K. (ed.), *Investing in Regional Development. Policies and Practices in EU Candidate Countries*. Budapest: Local Government and Public Reform Initiative, Open Society Institute, pp. 16–43.
Blažek, J. and Vozáb, J. (2004), 'The institutional and programming context of regional development.' In: Drbohlav, D., Kalvoda, J. and Voženílek, V. (eds), *Czech Geography at the Dawn of the Millenium*. Olomouc: Czech Geographic Society and Palacky University in Olomouc, pp. 255–266.
Čermák, Z. (2004), 'Migration aspects of suburbanisation in the Czech republic.' In: Drbohlav, D., Kalvoda, J. and Voženílek, V. (eds), *Czech Geography at the Dawn of the Millenium*. Olomouc: Czech Geographic Society and Palacky University in Olomouc, pp. 319–328.
Dostál, P. and Hampl, M. (2002), 'Metropolitan areas in transformation of regional organisation in the Czech Republic.' *Acta Universitatis Carolinae Geographica* XXXVII (2), pp. 133–155.
Drbohlav, D. and Sýkora, L. (1997), 'Gateway cities in the process of regional integration in Central and Eastern Europe: the case of Prague.' In: *Migration, Free*

Trade and Regional Integration in Central and Eastern Europe, Wien: Verlag Österreich, pp. 215–237.

Eskinasi, M. (1995), 'Changing housing policy and its consequences: the Prague case.' *Housing Studies* 10 (4), pp. 533–548.

Lahoda, et al. (1999), *General housing plan Brno*. City of Brno: Housing Department.

Lux, M. (ed.) (2003), *Housing Policy: An End or a New Beginning?* Budapest: Local Government and Public Reform Initiative, Open Society Institute.

Lux, M. (2004), 'Housing the poor in the Czech Republic: Prague, Brno, Ostrava.' In: Fearn, J. (ed.), *Too Poor to Move, Too Poor to Stay: A Report on Housing in the Czech Republic, Hungary and Serbia*. Budapest: Local Government and Public Reform Initiative, Open Society Institute, LGI Fellowships Series, pp. 25–66.

Maier, K. (1998), 'Czech planning in transition: assets and deficiencies.' *International Planning Studies* 3 (3), pp. 351–365.

Maier, K. (2000), 'The role of strategic planning in the development of Czech towns and regions.' *Planning Practice and Research* 15 (3), pp. 247–255.

Maier, K. (2002), 'The Prague Metropolitan Region.' In: Salet, W., Thornley, A. and Kreukels, A. (eds), *Metropolitan Governance and Spatial Planning*, London and New York: Spon Press.

MMR (2001), *Koncepce bytové politiky (Aktualizovaná verze Koncepce bytové politiky z října 1999) [Conception of housing policy (updated version)]*. Praha: Ministerstvo pro místní rozvoj.

MMR (1993), *Rozvoj měst a obcí: programy a dotační tituly MMR v roce 2003 [The development of cities and municipalities: programs and subsidies from the Ministry of Local and Regional Development]*. Praha: Ministerstvo pro místní rozvoj.

Ptáček, P., Létal, A. and Sweeney, S. (2003), 'An evaluation of physical and functional changes to the internal spatial structure of the historical centre of Olomouc, Czech Republic, 1980–2000.' *Moravian Geographical Reports* 11 (2), pp. 2–10.

Pucher, J. (1999), 'The transformation of urban transport in the Czech Republic, 1988–1998.' *Transport Policy* 6, pp. 225–236.

Rehnicer, R. (1997), 'New challenges for urban planning in Central and Eastern Europe.' In: Kovács, Z. and Wiessner, R. (eds), *Prozesse und Perspektiven der Stadtentwicklung in Ostmitteleuropa*. Münchener Geographische Hefte 76, pp. 63–71. Passau: L.I.S. Verlag.

Reiner, T.A. and Strong, A.L. (1995), 'Formation of land and housing markets in the Czech Republic.' *Journal of the American Planning Association* 61 (2), pp. 200–209.

Rietdorf, W., Liebmann H. and Schmigotzki B. (eds) (2001), *Further Development of Large Housing Estates in Central and Eastern Europe as Constituent Elements in a Balanced, Sustainable Settlement Structure and Urban Development*. Erkner: Institute for Regional Development and Structural Planning.

Strategic Plan for Prague (2000), Prague: City Development Authority of Prague.

Strong, A.L., Reiner, T.A. and Szyrmer, J. (1996), *Transformations in Land and Housing: Bulgaria, The Czech Republic and Poland*. New York: St. Martin's Press.

Sýkora, L. (1995), 'Prague.' In: J. Berry and S. McGreal (eds), *European Cities, Planning Systems and Property Markets*. London: E & FN Spon, pp. 321–344.

Sýkora, L. (1996), 'The Czech Republic.' In: Balchin, P. (ed.), *Housing Policy in Europe*. London: Routledge, pp. 272–288.

Sýkora, L. (1999), 'Local and regional planning and policy in East Central European transitional countries.' In: Hampl et al., *Geography of Societal Transformation in the Czech Republic*. Prague: Charles University, Department of Social Geography and Regional Development, pp. 135–179.

Sýkora, L. (1999), 'Processes of socio-spatial differentiation in post-communist Prague.' *Housing Studies* 14 (5), pp. 679–701.

Sýkora, L. (1999), 'Changes in the internal spatial structure of post-communist Prague.' *GeoJournal* 49 (1), pp. 79–89.

Sýkora, L. (2002), 'Global competition, sustainable development and civil society: three major challenges for contemporary urban governance and their reflection in local development practices in Prague.' *Acta Universitatis Carolinae Geographica* XXXVII (2), pp. 65–83.

Sýkora, L. (2003), 'Between the State and the market: Local government and housing in the Czech Republic.' In: Lux, M. (ed.), *Housing Policy: An End or a New Beginning?* Budapest: Local Government and Public Reform Initiative, Open Society Institute, pp. 47–116.

Sýkora, L. (2005a), 'Gentrification in postcommunist cities.' In: Atkinson, R. and Bridge, G. (eds), *Gentrification in a Global Context: The New Urban Colonialism*. London: Routledge, pp. 901–105.

Sýkora, L. (2005b), 'The Czech Republic.' In: Baan, A., van Kempen, R. and Vermeulen, M. (eds), *Urban Issues and Urban Policies in the New EU Countries*. Aldershot: Ashgate.

Sýkora, L., Kamenický, J. and Hauptmann, P. (2000), 'Changes in the spatial structure of Prague and Brno in the 1990s.' *Acta Universitatis Carolinae Geographica* XXXV (1), pp. 61–76.

Sýkora, L. and Ouředníček, M. (2005), 'Sprawling post-communist metropolis: commercial and residential suburbanisation in Prague and Brno, the Czech Republic.' In: Dijst, M., Razin, E. and Vazquez, C. (eds), *Employment Deconcentration in European Metropolitan Areas: Market Forces versus Planning Regulations*. Kluwer.

Surazska, W. and Blažek, J. (1996), 'Municipal budgets in Poland and the Czech Republic in the third year of reform.' *Environment and Planning C: Government and Policy* 14, 1, pp. 3–23.

Temelová, J. (2004), *Contemporary Buildings in City Promotion: Attributes and Foundation of High-Profile Structures. The case of Prague and Helsinki*. Research and Training Network Urban Europe.

Temelová, J. and Hrychová, H. (2004), 'Globalisation, eyes and urban space: visual perceptions of globalising Prague.' In: Eckardt, F. and Hassenpflug, D. (eds), *Urbanism and Globalisation*. Peter Lang, pp. 203–221.

Uhlíř, D. (2004), 'Regional versus national development: what sort of policy for the new Czech Regions?' In: Drbohlav, D., Kalvoda, J. and Voženílek, V. (eds), *Czech Geography at the Dawn of the Millenium*. Olomouc: Czech Geographic Society and Palacky University in Olomouc, pp. 269–277.

Chapter 9

The Third World in the First World: Development and Renewal Strategies for Rural Roma Ghettos in Slovakia

Jakob Hurrle

Introduction

Although ten years ago it was a topic hardly ever mentioned beyond the narrow circles of human right activists, the situation of the Roma minorities in Central and Eastern Europe has recently become the subject of various ambitious empirical studies and publications (European Commission 2002, Ivanov 2002, Ringold and Tracy 2002, Ringold et al. 2003, World Bank et al. 2002, Vašečka et al. 2003). The new interest of Western countries and international organizations reflected in these studies is a result of changes in Europe's political geography. With the Eastern enlargement of the European Union, the increasing social exclusion of Roma in countries such as Slovakia, Hungary or Romania has become a pan-European issue. This international dimension is the reason for the fear of the old member states that poverty, desperation, and open borders could result in a mass exodus to the West. In those countries in Central and Eastern Europe where Roma constitute a minority of significant size (such as Slovakia, where some estimate the Roma to constitute up to 8 or 10 percent of the population),[1] the Roma issue has been conceived as one of the main obstacles to their European integration. At the same time, however, the national governments used the Roma issue as an effective way to demand EU support to resolve a problem which might indeed exceed the economic and political capacities of the transition countries.

---

1   In the UNDP/ILO report, Andrey Ivanov refers to an estimate of the London-based Minority Rights Group, according to which there would be 480,000 – 520,000 Roma (nine to ten per cent of the total population) living in Slovakia. However, according to the recently published results of the 'socio-graphic mapping of Roma settlements', these numbers are probably too high. In the 'Atlas of Roma Settlements' published on the basis of this research, the researchers believe that the number of 320,000 (6.4 per cent of the total population) comes closest to reality (Radičová 2004: 11). Both estimates are much higher than the number of citizens (89,920 in 2001) which officially declare themselves to be of Roma origin (UNDP 2004: 15.) Regarding the difference between estimated numbers and official numbers see also: MG-S-ROM 2000: Roma and Statistics. Strasbourg: Council of Europe.

In most of the above-mentioned studies, one recurrent finding is direct the causality between poverty and spatial exclusion. The more spatially isolated from the majority population a Roma family lives, the higher the probability that this family lives in extreme poverty. Roma living in segregated settlements are also less able to maintain contacts with members of the majority population (Ivanov 2002: 70; Radičová 2004: 11). Obviously, this further limits their chances for a future integration into mainstream society. In light of the empirically proven correlation between poverty and spatial exclusion, the opening-up of existing ghettos and the prevention of any new ghettoization should be a cornerstone in any Roma integration strategy. Alas, current developments in Slovakia and other Central and Eastern European countries give little reason for optimism in this respect. *'Creation of living conditions for homeless citizens, citizens who failed to pay their rent or who are non-adaptable '*[2] – in many Slovak cities with a significant share of Roma this and many similar sounding municipal resolutions led to the creation of new ethnic ghettos for Roma who tended to occupy the historically valuable yet neglected historic inner cities during the socialist period. The newly created units of substandard social housing were typically created in spatial isolation from the rest of the city's housing areas. This way they are one of the most important stabilizing factors in the circle of social exclusion and poverty.

The creation of new ethnic ghettos can be conceived of as a reversal of developments during the socialist period, when the state sought to completely assimilate the Roma. Yet the liquidation of many traditional rural settlements and the dispersal of their inhabitants to urban areas did not only aim to remove spatial barriers, but was part of a state policy, which aimed consciously to liquidate the minority's culture, language, and ethnic identity (Barany 2002, Holomek 1998). Increasing conflicts between Roma and the rest of the population already signalised during the communist period that the policy of forced assimilation had failed. The creation of new ghettos after the end of the socialist period can also be understood as a delayed effect of this failed policy.

Due to housing shortages and a resistance towards the realization of the official assimilation policy on the part of both local authorities and the Roma themselves, the socialist state did not succeed in eliminating all rural Roma settlements. Depending on factors such as the policy of local authorities, geographic distance, and the relationships between majority and minority, over the years some of the former Roma settlements developed into integral parts of the majority villages while others remained underdeveloped and spatially segregated. In many locations,

---

2   The quotation is taken from Resolution Nr. 55/95 of the Košice city council. On the basis of this resolution, Roma living in the historic centre and other parts of the city were systematically forced to exchange their flats for substandard flats in Košice – Luník IX. At the same time, the 'non-problematic' part of the original inhabitants of Luník IX received exchange flats in the Žahaovce district. As a consequence of this policy, Luník IX is inhabited only by Roma today. The population density is the highest in all of Slovakia (also see Čapova 2003, Farnam 2003, Hurrle 1998).

unemployment among adults is close to hundred per cent. In many of the most underdeveloped settlements about half of the inhabitants are younger than 18 years (FAZ 2004).

The prospects for the development and renewal of those settlements belonging to the latter category are the subject of this paper. Ultimately intended as a contribution towards the development of strategies for the most segregated rural Roma ghettos in Slovakia, I start with the historic causes for the current marginalisation, followed by a description of the current situation, and a critical assessment of the strategies by which the Slovak government seeks to improve the situation in the most underdeveloped settlements. The article is based on the study of written sources, interviews with experts and Slovak policy-makers, and personal observations made during a number of research stays in Slovakia.

History

The Romská Osada as a Result of Empress Maria Theresa's Assimilation Policy

Since the time of their arrival in Europe in the Middle Ages, Roma have been experiencing rejection and persecution by the majority society. One of the few exceptions in this history of persecution was Hungary in the 17th and 18th century, which included also Upper Hungary, the territory now known as Slovakia. The relatively tolerant attitude of many Hungarian lords towards the Roma caused a migration of Roma from Western Europe. As a consequence of this migration, to this day a large share of the European Roma lives in Hungary and former Hungarian territories.

Yet the tolerant attitude of the Hungarian nobility also caused conflicts with the rulers of the Habsburg empire, who had gained a reputation of being extremely fierce advocates of 'vigorous anti-Gypsy measurements.' In the early years of her reign the Austrian empress Maria Theresa (1740–1780) continued the hostile policies of her predecessors. In 1758, however, the issuing of her first decree regarding the question of the Gypsies marked the beginning of a new policy, aimed at immobilizing and assimilating the Roma. This turn in the state's policy towards the Roma was inspired by the ideas of the Enlightenment. In difference to previous times, the otherness of Roma was no longer explained as the result of an inherited defect, but rather as the correctable consequence of inferior socialization (Frazer 1995: 156–9).

According to the ideas of Maria Theresa, the Roma should be completely assimilated into the peasantry. Yet the empress's first decree, which dealt with the settling of Roma and their obligation to pay taxes and deliver labour for their lords, provoked resistance in the villages where Roma were to be settled.

One reason for controversy was the explicit wish of the empress that Roma should build their houses *'in linea'* with the houses of the majority population:

Die Zigeunerhäuser, welche noch nicht in der Ordnung der anderen Häuser aufgebaut sind, auch wenn sie von gutem, harten Zeug erbaut sind, müssen alle abgerissen und zusammengeworfen und in der Reihe und Ordnung der anderen Häuser erbaut werden, und alle sollen ihre Contributionszahlungen entrichten.[3]

However, not only the resistance of the peasants against the integration of Roma into their community, but also the opposition on the part of the Roma, was stronger than expected in Vienna. Some of the Roma, whose official designation was changed in 1761 by another decree to *Ujmagyar* (New Hungarians), left their new houses in midst of the villages of the *gadže* (as Non-Roma are called in the Romany language) to either continue an itinerant life style or to live with their own people in self-built shacks outside the villages of the majority.

Despite such setbacks, the assimilation policies of empress Maria Theresa and her son Joseph II led to the sedentarization of most Roma in Austria-Hungary. This would make their life style different from the one of the Sinti and Roma in Western Europe, which in their majority continued to live a nomadic life long into the 20th century. To this day one can find Roma settlements (in Slovak called *ciganská osada* or *romská osada*) On the outer margins of many villages in Southeast Poland, Hungary, Slovakia, and the Carpatho-Ukraine and many of them can be traced back to the days of the assimilation policy of Maria Theresa and her son Josef II.

Modernization

Modernization as Loss of Economic Niches One of the objectives of the Austrian assimilation policy was to disaccustom the Gypsies of their typical occupations. However, at the beginning of the 20th century, most of the Roma in Austria-Hungary continued to make their living thanks to the same economic niches Roma had been occupying since their arrival in Europe. In the case of Slovakia, the most important economic activities for Roma were the production of wrought-iron work and musical performances for the local population. None of these occupations required the ability to write or read or any other formal education. Despite the rise of industrialized production, during the 19th century these traditional occupations were still sufficient to maintain a decent standard of living in the economically underdeveloped Hungarian part of the empire, meaning the Roma were not much worse off than the rest of the rural population (Barany 2002: 14). At the same time, their economic usefulness secured the Roma a certain respect in the eyes of the majority population who may not have valued the presence of the Gypsies as such but made use of their services.

Due to the enormous socio-economic gap between the industrial centres of Western Europe and the continent's Eastern half, most of Eastern Europe remained dominated by agriculture until the mid of the 20th century. Despite this

---

3   'The Gypsy houses, which are not yet built in the order of the other houses should all be destroyed, even if they are made out of good, hard material. They should be rebuilt in line with the other houses and everyone should pay taxes.' Order from the year 1774, quoted after Mayerhofer (1987: 172).

underdevelopment, at the beginning of the 20th century the effects of modernization were already sufficient to start destroying the economic basis of the relationship between the farmers in Eastern Europe and the their Roma neighbours in Upper Hungary (Kollárova 1992, Bartoloměj 1998). One consequence of this development was increased conflict between farmers and Roma, reinforced by the increasingly difficult economic situation of the small peasantry and the natural increase of the Roma population. During the interwar-period, these conflicts had only limited a impact on the Gypsy policies of the Czechoslovak state, which mainly focused on the control of those Roma and Sinti who were still living a nomadic life. However, at the local level, the increasing conflicts between sedentary Roma and the rural population led to violent acts and pogroms in some cases (Kollárova 1991: 64, for a comparision with the developments in the Austrian Burgenland see Mayerhofer 1987: 36–43).

Modernization as Systematisation of Persecution In the occupied Czech part the definitive defeat of Czechoslovakia led to the establishment of the German protectorate Bohemia-Moravia in 1939, while Slovakia became a formally independent country by the grace of Germany, governed by the populist party of the catholic priest Jozef Tiso. In the protectorate and in the Slovak puppet state alike, the German influence led to the introduction of racial laws which targeted both Jews and Roma. In Bohemia and Moravia, about 90 per cent of the Roma population were murdered in concentration camps. In difference to the situation in the protectorate, the clerical-fascist regime of Slovakia proved less determined in the persecution of Roma than in the 'solution of the Jewish question.' Historic documents indicate that Bratislava was preparing for the extermination of the country's relatively large Roma population (the historian Ctibor Nečas estimates 100,000 people). However, local party functionaries demanding concrete actions in this direction were told to be patient, since the state would have had difficulties replacing the (forced) labour the Roma supplied under the conditions of a war economy. For the time being, the state pursued a policy of exclusion and economic exploitation. One example for this policy was the prohibition to enter the centres of larger cities and the health resorts of the High Tatra mountains. Roma were also incarcerated into labour camps at a large scale. In addition to this, in 1943 the government issued a decree that the existing Roma settlements in proximity to highways had to be removed to isolated locations (Nečas 1992). In many cases, the transferred settlements remained in their new location after 1945. To this day distances of several kilometres between some Slovak villages and their *osady* are a direct consequence of the persecution in the time of the Slovak state (Jurová 2002).

Modernization as Forced Assimilation In 1948 the communist rise to power led to another paradigm change in Roma policy in Czechoslovakia. The fascist ideology of racial supremacy, according to which Roma were seen as genetically inferior beings, was replaced by an ideology which explained the 'cultural backwardness' of the Roma as a relic of feudal society.

The first systematic step towards a Roma integration policy was a decision taken in 1952, according to which the Roma would be subjected to a 're-education' programme. Re-education meant that the Roma were expected to give up their cultural roots and become good socialist Czechs or Slovaks. Included in the programme were measures forcing the Roma into new labour relations in the socialist economy. Roma were no longer allowed to pursue their traditional occupations or to make their living as self-employed musicians (Jurová 2002).

The re-education programme proved much less successful than had been expected by its initiators. This was mainly caused by the fact, that the programme had been designed without any knowledge of the culture and mentality of the Roma, let alone any consultation with their representatives (Holomek 1998: 3). The programme's disappointing results spurred a longer discussion within the party nomenclature. Despite increasing conflicts, one group of decision-makers pleaded for an intensification of the policy of forced assimilation. The alternative proposal was to recognize the Roma as a national minority just like ethnic Hungarians, Ukrainians and Germans, all of which would be given the right to preserve their distinct cultures and languages. In 1958 the Communist Party's Central Committee decided the controversy in favour of the first group's ideas. In consequence, the state planned to eliminate all of the *romské osady* in Slovakia, which were at this point the home of about 10,000 families (63,000 persons) (Holomek 1998: 4). In addition to this, the usage of the Roma language was forbidden in the public sphere. Children from Roma families who used their mother tongue at school were punished, and until the 1980s even the performance of Gypsy folklore was forbidden (Mann 1992).

Due to lacking housing capacities and the steadily growing Roma population, the communist state failed to realize its planned dispersal of all Roma from rural settlements over the whole territory of the ČSSR. The forced assimilation policy, conducted without any consideration for the culture and mentality of the people concerned, did nevertheless cause irreparable damage. One of the reasons for the conflicts between Roma and the majority population was the forceful destruction of the traditional hierarchies and the accompanying loss of traditional authorities, which had traditionally functioned as linking elements between the Roma and the majority Holomek 1998: 4).

The Prague Spring marked the end of the most inhumane stage of the Czechoslovak assimilation policy. For a short period, Roma were even given the chance to organize in cultural organisations. In the course of the 'normalization' period following the Soviet invasion, the policy paradigm shifted once more towards assimilation. However, it neither reached the vigour, nor the grade of repression of the years before 1968. The decreasing zeal on the side of the state authorities was also a consequence of the increasing problems in the relationship between Roma and the majority. To the rulers of the socialist state, the evident failure of their policy constituted a serious dilemma:

The whole thing is very embarrassing for the socialist countries. They explain the 'Gypsy way of life' as a consequence of [historic] class exploitation. Yet this makes it difficult to explain the stability and further spreading of this way of life [under the condition of a socialist society].   (Gronemeyer 1983: 211)

Trying to avoid further escalation, the state silently reversed its aim to achieve the assimilation objective in the 1970s and 1980s. This decision was accompanied by the launch of an intensive social policy. Accordingly, the problem group became the target of the intensive care by the state's social institutions and enjoyed special benefits. However, in the opinion of Roma activist and former dissident Karel Holomek, this intensive social care had a negative impact on the Roma, increasingly damaging their independence and self-reliance:

[The Roma] became accustomed to this safety net. They would also use the health and educational welfare system to put their children in hospital [sic] from time to time or send them to children [sic] homes on a permanent basis.   (Holomek 1998: 4)

The Roma in Times of Crisis   For the reasons given above, the Czechoslovak communist policy towards the Roma deserves an ambivalent assessment at best. Despite of this, most ordinary Roma remember this era with nostalgia. In their view, relative material wealth, a great degree of material security, abundance of labour, and the protection from physical violence were all important factors making the communist regime superior both to preceding times of persecution and the period since 1989 (Barany 2002: 151–3).

Today, Roma suffer more than most other groups from the breakdown of the socialist economy. Due to their lack of other qualifications, Roma worked mainly in sectors demanding large amounts of manual labour, such as mining, heavy industry, forestry and farming. In socialist times, when state ideology favoured industrial production while neglecting the service sector, these were industries of great relevance. Post-socialist rationalization and de-industrialization disproportionately affected employment in these and other industries in which Roma had found their new economic niches.

Roma were of course not the only group that experienced material insecurity and a loss of social status as a consequence of economic reform. Yet the Roma, more than most others, lacked education and resources like property or access to professional networks, which helped other groups to cope with the new situation. As a result of the assimilationist policies of the socialist period, but also because of the general difficulty to maintain traditional structures in an increasingly modern environment, the traditional organizational structures within the Roma minority were no longer strong enough to balance their lack of social integration. Accordingly, Roma are much more dependent on state welfare than most other groups. Independent of racial discrimination, the described developments have to be seen as the key reasons for the increasing marginalisation of the Roma after 1989.

The Slovak Roma After 1989 – The Spatial Dimension of Marginality

Spatial Distribution of the Slovak Roma

Despite communist attempts to disperse the Roma over the entire territory of Czechoslovakia, about two-thirds of the Slovak Roma continue to live in the Eastern and Southern parts of Slovakia, which have traditionally had the highest concentration of Roma. At the same time, these are also the economically most disadvantaged regions of Slovakia. Especially outside the larger cities, unemployment rates are very high (Džambazovič and Jurásková 2003). The Slovak sociologist Iveta Radičova coined the term 'double marginalisation' to describe the fact that the Roma of the Eastern Slovak settlements are marginalized not only in relation to their Slovak neighbours, but also (together with these neighbours) in relation to regions with better economic prospects (Radičova 2003). Especially the inhabitants of separate settlements in rural areas, accounting for an estimated 25 per cent of the total Roma population in Slovakia, are in an economically hopeless situation (Džambazovič and Jurásková 2003: 332). The demand for unskilled labour in agriculture, forestry and construction decreased dramatically as a result of the economic transformation. Being traditionally landless, Roma also did not gain from the re-privatisation of agricultural land that followed the break-up of many collective farms. In consequence, unlike most other people in the impoverished rural areas of Eastern Europe, Roma have limited possibilities to use subsistence agriculture as a survival strategy (Ivanov 2002: 16).

Spatial Isolation and Poverty

Comparative surveys on Roma living in different surroundings show a clear relation between the extent of spatial isolation / integration and their participation in the economic and societal life beyond the limits of their ethnic group. According to a UNDP/ILO survey, a majority of those respondents who declared to have no contact at all with the majority population (17 per cent) lives in isolated rural settlements (Ivanov 2002: 70). Roma living in this type of settlement also have the highest number of children (on average 7,8 according to the UNDP/ILO report in 1997). They also answered 'yes' to the question *'Were there periods during the last year when your family did not have enough to eat?'* much more often than urban Roma (Ivanov 2002: 45). A World Bank study on poverty among Slovak Roma describes the relation between living conditions and geographic location in the following words:

> Within regions, the level of poverty in a Roma settlement appears to be closely connected to the geographic location of the settlement, and the level of ethnic integration and segregation. Conditions in settlements which consisted only of Roma were significantly worse off than conditions in more integrated communities.   (World Bank et al. 2002: 12)

Despite of these empirical findings it is extremely difficult to make generalization about the character of the various types of settlements. One reason for this is a lack of data and difficulties in establishing clear criteria and definitions, another is the Roma minority's extreme heterogeneity. As will be shown in the following, this applies especially for the situation in rural settlements.

The Slovak Romská Osada in the 21st Century

Types of Settlements When reporting on the situation of Roma in Slovakia, journalists tend to focus on infamous places like Letanovce, Rudňany, Jarovnice or Svinia, where living conditions are among the worst in all of Europe. Yet this one-sided focus leads to a distorted perception of a much more complex reality. According to the recently published results of the first comprehensive mapping of Roma communities in Slovakia, about half of the estimated 320,000 Slovak Roma lives integrated among the majority population. Out of 1575 places identified by the majority population as 'inhabited by Roma', 338 with 64,661 inhabitants are located on the margins of villages and town, 281 with 49,586 inhabitants are separated from the rest of the built environment by their distant location or a spatial barrier (Radičová 2004: 11).

However, if using the definition of segregation developed by the authors of the above-mentioned study,[4] only 149 out of these 619 settlements on the margins or outside of the built environment qualify as segregated settlements. Using an even narrower definition of segregation and underdevelopment ('no or almost no infrastructure'), the study defines a group of 46 settlements, which lack running water, sewage and gas, and are not accessible via an asphalted road. The quality of housing is extremely poor in the most segregated settlements but not only there. Of the settlements located beyond the borders of villages and towns, 21.2 per cent live in tiny wooden shacks that in most cases are the home of entire families with an average occupancy of 6.2 persons (Radičová 2004: 11–21).

Types of Construction In his description of various types of housing structures in Slovak Roma settlements the cultural anthropologist Alexander Mušinka from Prešov University points to the fact that it is difficult to make generalization about the overall situation on the basis of descriptions of individual settlements. In reality, many settlements could be categorized somewhere in between two pairs of extreme poles: *integrated – segregated* and *traditional – modern* (in relation to the architecture). The extreme differences between the settlements would make it difficult to assess the built structures objectively, since 'what is considered a shack in the Roma settlement in the village of L´ubotín would be an above-standard house in the Roma settlement in Svinia' (Mušinka 2002: 327).

---

4   The authors define a segregated settlement as 'being located on the margins or outside of a village/town, having no access to municipal water-supply and more than 20 per cent of illegally erected buildings.' Radičová, Iveta (ed.) 2004, 11.

Figure 9.1   Different types of Roma settlements

In previous times, the most common building structure was a tiny house with only one room, which was usually not larger than 3.5 square meters. Built without fundaments, such buildings are still a common feature in the least developed settlements. Roma were already dependent on technical assistance or building materials from the majority population in the past. Hence, the influx of modern construction methods also led to a 'modernization' of the *osada*. One example for this is the replacement of traditional materials like wood or clay bricks by concrete panels or the usage of bricks from razed houses in the villages of the majority.

As mentioned earlier, the communist government did not succeed in liquidating all traditional Roma settlements in Slovakia. As a result, some of the existing settlements were modernized rather than liquidated. Especially in places where the Roma settlement was immediately adjacent to the majority village, some of the Roma settlements gradually developed into normal parts of the village. The family houses erected as a result of such modernization attempts do not differ much from other buildings erected during the socialist period. In other places, local authorities aimed to modernize the Roma settlement without eliminating the spatial border between Roma and the majority. One typical result of such policies is single multi-family blocks made out of prefabricated concrete panels within the existing *osada*. Often, however, the modernization of the building stock did not keep up with the natural growth of population. As a consequence, in some places a 'traditional osada' became surrounded by the structures of a 'modern osada'. Post-socialist transition brought an end to the construction of new social housing. Meanwhile, the steady increase in Roma population and the poverty-driven back-migration of urban Roma to the countryside caused the existing ghettos to grow. In a growing number of localities, Roma begin to outnumber the number of ethnic Slovaks/Hungarians living in the village of the 'majority'. Yet due to a widespread lack of political unity, in most cases Roma were not able to make political use of these demographic changes. Another effect of the growth of the Roma population is the rise of new segregated settlements, for example in abandoned factories or mines.

'Enclaves of Traditional Society' vs. 'Culture of Poverty'

The described differences in the built structures mirror the differences in social structures. This applies to the relationship between Roma and the members of the majority population, but also to the conditions within the local community. The importance of the latter is well illustrated in an article by Czech anthropologist Marek Jakoubek and in a directly related piece by his colleague Karel A. Novák. The subject of both articles is the theoretic conceptualisation of the phenomenon *'romská osada'*.

In the first piece, Marek Jakoubek argues for an understanding of the *osada* as one of the 'last enclaves of traditional society' in the Western world. In difference to the surrounding world, social structures within the rural Roma communities are not based on the egalitarian citizen principle of modern Western societies, but on the pre-modern principle of family relationship and the delimitation from the world beyond the own *osada.*(Jakoubek 2003: 26).

Karel A. Novák accepts the possibility that there might still exist *osady* functioning on the basis of those traditional principles described by Jakoubek.[5] However, in his opinion, life in the majority of rural settlements does not differ fundamentally from the life in the modern Roma ghettos, which have been appearing in many Czech and Slovak cities since the beginning of the transformation. In 'modern' urban and 'traditional' rural ghettos alike, life is dominated by a self-perpetuating *culture of poverty.* Novák accepts the central role of family relationships, yet contests Jakoubek's interpretation of them as networks of solidarity: 'It is no longer possible to speak of family structures as a basis for solidarity, since under the conditions of the 'culture of poverty' there does not exist any stable solidarity (Novák 2002: 38).

The term 'culture of poverty' used by Karel A. Novák was initially developed by the American anthropologist Oscar Lewis, who first used the concept in his 1966 study of a family from Puerto Rico (Lewis 1966: 19–25). According to Lewis, a 'culture of poverty' is defined by a self- perpetuating set of attitudes and behavioural patterns which exist independently from 'poverty as such' and therefore need to be studied as a separate phenomenon. In the Central and Eastern European debate on the Roma issue, Karel A. Novák is not the only one for whom Lewis' thesis seems to provide a suitable conceptual frame for the problems of the *romská osada*:

> Aside of [sic] ... the lack of social organisation und the diminution of solidarity to the nuclear family there are the following attributes: A life strategy oriented on the presence [sic], lack of private property, and a closed economic system, characterized by the sale of private goods and the lending of money at high interest rates. The members of these communities have great mistrust towards the outside word. ...Most important are the consequences for the children, which grow up in this milieu...Such a family raises children dominated by feelings of fatalism, powerlessness, dependence and inferiority.

In the *romská osada* we find almost all of the phenomena described by Oscar Lewis. Most important of all is probably the lending of money at high interest rates. This phenomenon is spread [sic] in most of the settlements and involves the majority of their inhabitants. However, there exist also *osady* in which this phenomenon doesn't exist at all (Novák 2002: 39).

Not only Novak's description of the reality, but also in his sceptic view of the possibilities and limits of traditional social policy with an emphasis on re-distributive measures is clearly inspired by the Anglo-Saxon debate on ghettos and the *ghetto poor*. He quotes the American social scientist Edward Banfield as saying that there are two categories of poverty:

---

5   Karel A. Novák poses the question in which way such settlements would differ from traditional Slovak villages, which could also be seen as 'enclaves of traditional society' (Novák 2003: 37).

'Some have simply no money but adhere to the values of the middle class. They can therefore profit from support by the state.' The second group is the real lower class. Oriented in their life strategies strictly on the present, these people 'would continue to live in poverty even if their salaries would double'... (Novák 2002: 38)[6]

Opening up the Ghetto

The increasing international awareness of the situation of the Roma in Central and Eastern Europe brought many private donors and international organisations to Slovakia aiming to improve the minority's situation through various projects. Examples for activities targeting segregated settlements in Slovakia are the training and employment of social field workers (Sandor Marai Foundation, Milan Šimečka Foundation, ETP, PHARE, and others), the establishment of community centres and kindergartens (PHARE, UNDP/MATRA project 'Your Spiš', Spolu, and others), and investments in technical infrastructure (PHARE) and housing (Habitat for Humanity, SPOLU).

Despite the increasing volume of resources invested, the projects did not succeed in reversing the general trend towards an increasing exclusion of the Roma minority. On the local level, however, some of the mentioned projects and organizations succeeded in making a difference. Together with interested social scientists and sympathetic state officials, the people working for these projects were also successful in lobbying for changes in government policy. Despite these changes, Slovakia is still far from having a systematic strategy for the inclusion of her Roma citizens. Yet in case that there will once be the political will to formulate a truly coherent strategy, it will for sure be built on the know-how and human capacities built during the realization of projects such as the ones mentioned above.

Towards a Systematic Government Strategy

The situation of the Roma population was one of the most difficult issues in the membership negotiations between Slovakia and the European Union. Accordingly, the Slovak government was forced to demonstrate greater political will to resolve this issue than in the politically complicated first years after the breaking up of Czechoslovakia in 1993 (Sobotka 2001). The government created the position of the 'plenipotentiary of the Slovak government for Roma communities', who is to 'realize the systematic efforts for the solution of issues concerning the Roma communities, the improvement of their status and their integration into the society.'[7]

---

6   The quotations are re-translated from Czech and refer to: Banfield, E.C. 1970: The Unheavenly City. The Nature and Future of Our Urban Crises. Boston.

7   Štatút splnomocnenca vlády Slovenskej republiky pre rómske komunity (vládou SR schválený 17. decembra 2003, uznesenie č. 1196/2003) [Statue of the Slovak Government plenipotentiary for the Roma minority, Decree on Dec. 17th, 2004, No. 1196/2003], translation JH, Paragraph 3, Abs. 1.

In addition to this, the government decided upon an overall strategy for the inclusion of Roma in Slovakia. In the eyes of many, however, this official government strategy is barely more than a non-binding declaration of vague goals. Critics also point to the weak position of the government plenipotentiary, who has neither significant budget resources on her own, nor a clear competence to intervene into the activities of other ministries. As a consequence, critics lament that the state's activities are fragmented into many, seemingly unconnected activities.[8]

Another frequently point of criticism is that concrete activities are usually limited in size and often financed by external donors (Sobotka 2001). In the course of the last years, a large percentage of this external money came from the PHARE programme, which financed a number of education and infrastructure projects. In the opinion of many observers, the impact of most PHARE projects has been quite limited (Hriczko and Magdolenova 2003). This assessment is supported by a number of evaluation reports conducted by an independent consultancy, which criticize considerable delays in the realization of the projects and the lack of a government strategy to secure the sustainability of project outcomes.[9]

One often overlooked result of PHARE and various other Roma-targeting programmes is the growing number of people who are professionally involved with Roma issues. Many Roma see this development of a 'Gypsy industry' critically and accuse social workers, teachers and social scientists of making a comfortable living from their poverty. In a number of cases such accusations are justified. However, the gradually emerging professional networks of people and organizations involved in the field of Roma-targeted social policy has been important in slowly pushing the government towards a more systematic approach. Despite the existing flaws in the government's Roma policy, organisations such as the 'Association of Social Field Workers' were successful in lobbying for a number of legislative changes and the creation of new programmes, which might be seen as first cornerstones in a gradually emerging strategy for the social inclusion of Roma.

Social and Community Work

The negative experience with slum demolition programmes in the communist era shows that investments into technical infrastructure and housing alone are not sufficient to improve the situation of Roma living in segregated settlements. One precondition for a sustainable improvement is the preparation of the people who are to move into the new houses: Social work and community mobilization is necessary in order to give the beneficiaries of the technical investments the chance to participate in the changes concerning their settlements. Without such participation, it is likely

---

8   For a critical assessment of the Slovak government strategy see: Scheffel (2004), Guy (2001: 285–323), Zoon (2001), and Sobotka (2001). Regarding the office of the plenipotentiary, see: Vašečka (2002).

9   The independent evaluation reports can be found on the website of the Slovak government (http://www.vlada.gov.sk/mensiny/projekty_phare.html).

that the future inhabitants will not identify with their upgraded settlement. Social workers might also be needed as moderators in cases of conflict, which might for example arise when the demand for new housing exceeds the projected capacities. The need to prepare slum upgrading with social work is also recognized in the plenipotentiary's 'Comprehensive Development Programme for Roma settlements', which became official state policy in 2002.

There have already been a number of programmes in Slovakia which trained Roma and Non-Roma to work as social and community workers in segregated communities.[10] Until recently, however, it was not possible to pay social field workers out of the regular state budget, because the Slovak social legislation did not provide such a position. Consequently, many trained social workers had difficulties in finding employment, while successful projects ended with the end of external funding. A lack of sustainability has also been an issue in the case of some of the community centres which were established by private initiatives and the PHARE programme.

Over the course of the past years, the government finally incorporated the position of social field workers into the legal framework for social policy. As a first step in this direction, the office of the plenipotentiary started its own pilot project in 2002. Beginning in 2005, municipalities can apply directly with the ministry for social affairs to finance the employment of social workers. While the pilot programme started with around 22 field workers and 3 co-ordinators, the current state budget contains resources for up to 200 community workers, 400 assistants, and 20 co-ordinators. The future will show whether the ministry will be able to enlarge the programme at this pace without giving up on issues as quality of personal, professional trainings, and monitoring.

Technical Infrastructure and Housing

According to the results of the socio-graphic mapping, 45 percent of the buildings in settlements on locations outside of villages were built illegally (Radičová 2004: 16). In many cases, it is not even clear who the real owner of the land is. (Regarding the complicated land ownership issue, see Jurová 2002). Any legal investment into technical structures hence requires a legalization of the settlement and its incorporation into the villages' land use plan. This gives the municipality significant power to obstruct any unwelcome investment. Under these circumstances, it has proven almost impossible to demand from the local municipality, that the upgrading of housing should be accompanied by a reduction of the spatial distance between the Roma settlement and the village. In many cases, local authorities justify their opposition to a less distant location

---

10 An overview on Roma-related activities of various organisations in Central and Eastern Europe can be found in the UNDP 'Experiences Portal' (http://roma.undp.sk) and in the 'Innovative Practices Database' of the Open Society Institute's Local Government Initiative (http://lgi.osi.hu/ethnic/csdb/).

with the argument that no owner of lots in such locations would be willing to sell land for this purpose. In some cases, the planned renewal of the settlement will even increase the spatial distance between Roma and the majority.

The government's official strategy for the development of segregated Roma settlements is the 'Comprehensive Development Programme for Roma settlements'. The office of the plenipotentiary selected about twenty municipalities, where local authorities are willing to co-operate. In a first stage, the programme seeks to support the municipalities in the clarification of the land ownership issue. The employment of social field workers is another measure to prepare the ground for future investments in housing and technical infrastructure. However, it is symptomatic for the weak position of the plenipotentiary that the programme passed without the allocation of adequate funds. To realize the programme in its planned extend, the office is trying to acquire external funding. Meanwhile, participating municipalities have the possibility to accelerate the realization of the project objectives by applying on their own at the relevant ministries for government grants.

One important grant programme is the social housing scheme of the Ministry of Construction. This programme assists municipalities in the construction of low-standard social housing for inhabitants of rural Roma settlements. Together with PHARE project SR 0103.02, which finances the creation of technical infrastructure in twenty-nine rural settlements, the housing programme is thought to significantly improve the living conditions in some of the least developed settlements. As all the other PHARE programmes, project SR 0103.02 was also not developed in the office of the plenipotentiary, but by the Section for Human Rights in the Slovak government office. Accordingly, the twenty-nine targeted municipalities are not identical with the ones involved in the plenipotentiary's 'comprehensive development programme'.

However, even municipalities not participating in one of the two comprehensive programmes can apply for the social housing programme mentioned above. To be eligible for a grant, which covers eighty per cent of the construction costs, municipalities with socially marginalized citizens need to contribute municipal land with access to technical infrastructure and issue a building permit. The future inhabitants of the buildings are required to contribute the remaining twenty per cent of the construction costs, usually in the form of labour. Especially when prepared and accompanied by social and community work, this practical participation is likely to increase the identification of the inhabitants with their new homes. The visible contribution of the Roma might also increase the programme's acceptance on the part of local non-Roma, who tend to resent any programmes explicitly targeting Roma. However, the participation of the future residents in the construction is not at all popular by most construction companies, who have to account for the fact that the future residents often lack skills and work experience.

Work

According to empirical studies, most Roma see the lack of employment opportunities as their most serious problem (Ivanov 2002: 31). Paving access roads and the construction of new substandard housing will achieve not much more than veil the prevailing exclusion, if it is not linked with the generation of employment possibilities. Considering most Roma's poor educational background and the extremely high unemployment rates in those regions where most Roma live, it seems unlikely that the growth of the private sector will be able to change this situation much.

In order to at least provide people with temporary employment, Slovakia started to develop public work programmes. Due to the drastic cuts in social welfare, there now exists a huge demand to earn an additional 1,500 Slovak crowns (around 40 €) by taking part in the 'activation' programmes that replaced the former public work schemes last year . The cost of this programme is mainly financed by the European Social Fund. At the local level, the work is administrated either by NGOs or the local municipality. Most of the mayors interviewed for my research in principle welcomed the idea that people should receive welfare payments only if they deliver services in return. However, they criticized that the recipient's very limited work duty (ten hours a week) would not allow for serious work or an investment into their skills. In consequence, most of the Roma working in this scheme are used to clean public spaces.

The Crucial Role of Local Authorities

One important lesson of most projects targeting Roma concerns the crucial role of the municipality. In some cases, this crucial importance became apparent in a very negative way. Due to the informal status of many settlements, the local authorities have many possibilities to obstruct a project. Sometimes, this has happened, because the local major felt disrespected by a private initiative. In other cases, however, the obstruction is motivated by a general disapproval of projects in favour of Roma. The 'classic' example for the latter case is the municipality of Svinia, where a number of initiatives had worked for years to improve the conditions in the local Roma settlement. Elected after a populist anti-Gypsy campaign, a new major stopped the realization of a project's infrastructure elements financed by the European Union. The opponents could not be persuaded by the argument that the project would not only improve the disastrous living conditions in the *osada*, but also connect most of the majority village to an urgently needed sewage system.[11] According to plenipotentiary Klára Orgovánova, Svinia is not a unique case:

---

11 For comprehensive information on the Svinia settlement see the articles by David Scheffel, a Canadian anthropologist who belonged to the initiators of the 'Svinia Projekt': Scheffel (2004) and Scheffel (1999).

Many municipalities hope that the Roma will go one day, if the conditions get worse and worse. We try to convince them: The Roma will not go. And the problems will not disappear if the municipalities are not interested in solving them.[12]

However, in cases where activities gain the support of the local major, the local municipality's involvement is likely to greatly increase a project's sustainability. A very successful project is the UNDP/MATRA project 'Your Spiš'. One task of the Your Spiš community activists was social work, another was the interaction and co-operation with the local major. Especially in small municipalities, the project convinced a number of majors that 'Roma projects' can be designed in a way that not only the Roma but the entire municipality gain from it.

Most outside initiatives aimed at initiating local change are designed to operate only for a limited time. Unlike people and organisations coming from outside (who will leave one day for another project) or local NGOs (who can dissolve or run out of money), the municipality belongs to the few institutions that will always exist in the village. In many cases, this makes a municipality with a supportive major the best institution to secure that a project will continue after its (external) initiators have gone.

Due to the experiences summarized above, almost all newer initiatives choose to work in municipalities where local authorities are supportive. This is true not only for private initiatives, but also for most of the government programmes (PHARE infrastructure, comprehensive development programme, etc) which selected municipalities not purely on the basis of greatest need, but also of the local authorities' readiness to co-operate. From the perspective of project effectiveness, such a pragmatic approach is certainly reasonable. However, it seems nevertheless problematic that those Roma which would most need external support are not reached at all. Yet solving this dilemma would require legislative changes which would limit municipal autonomy. One possible approach could be the introduction of provisions in planning laws which would *force* municipalities to act (and respectively empower the state to intervene) if the housing conditions endanger the health of the local citizens. One could also envision a legal act which would automatically legalize Roma settlements that already existed during the communist period. However, anyone aiming to restrict municipal autonomy will confront fierce resistance. Considering the unpopularity of any programmes reminding people of the alleged 'positive discrimination' of Roma during the communist era, it is clear how difficult and unpopular a task it would be to push through such changes.

Conclusion

The integration of Roma into Slovak society represents a task of unique complexity. Unlike in the case of many other political challenges, this complexity is based less on a lack of material resources than on the difficulty to generate public support

---

12   Quoted on the bases of her contribution to a workshop of *Partners for Democratic Change*, held in Bratislava in 2004, June 16.

for activities in favour of an unpopular group. Part of this problem is the lack of political will on the part of the national government. Another part is the attitude of many majors and local councils which openly declare their determination to obstruct development projects that target the Roma in their municipality. However, the lack of development partners at the local level is not limited to the side of the majority population. It also concerns the Roma community which does not have many respected leaders representing the interests of their constituencies and functioning as counterparts and partners of the local authorities.

Due to these and many other difficulties, the Roma issue is often presented as being 'unsolvable'. Such a fatalistic view should be rejected, however. There have been quite a number of local-level projects which were successful in initiating positive developments. Many would have produced even better results if they were embedded within a coherent political strategy. It is mainly because of a lack of co-ordination and continuity that all the money flowing into Roma-targeted projects seems to have achieved little in reversing a general trend towards increasing exclusion. One address for this critique are independent donors, which were often not patient enough to realize that achieving real changes might be not a question of years, but of generations. However, the first address for criticism should be the Slovak government which still lacks the political will to pursue a real programme for the inclusion of the Roma.

Thanks to the experiences of many small and medium-seized projects, it is fairly clear what a comprehensive development strategy for the most segregated and underdeveloped settlements would have to look like. The central idea is the combination of social work and community mobilisation, vocational trainings and job creation, with the gradual upgrading of technical infrastructure and housing. At least on paper, the government already adopted these ideas back in 2002 with the passing of the 'comprehensive development programme for Roma settlements.'

However, it is necessary to be realistic about one thing. A renewal of the worst settlements in the outlined way will not be sufficient to overcome the Roma minority's increasing spatial isolation. In some cases, the modern settlements are planned in even more remote locations than their underdeveloped predecessors. In such cases, the new settlement might appear less like a slum, but will continue to bear the features of a ghetto. In the best case, the technical renewal of segregated settlements can be the first step towards a real integration of the inhabitants. However, it seems questionable that it is possible to stop segregation processes in a democratic country as long as most members of the majority population prefer to not have Roma as neighbours.

References

Barany, Z. (2002), *The East European Gypsies. Regime Change, Marginality, and Ethnopolitics*. Cambridge: University of Cambridge Press.

Čápova, H. (2003), 'Slovakia's Roma Knocking at the Door.' In: *RESPEKT* 50/2003, Prague.

Daniel, B. (1998), *Geschichte der Roma in Böhmen, Mähren und der Slowakei* [History of the Roma in Bohemia, Moravia and Slovakia]. Frankfurt am Main: Peter Lang.

Džambazovič, R., Jurásková, M. (2003), 'Social Exclusion of the Roma in Slovakia.' In: Vašečka et al., pp. 325–356.

European Commission (2002), *EU support for Roma communities in Central and Eastern Europe*. Brussels: Directorate General for Enlargement, Enlargement Information Unit.

Farnam, Arie (2003), 'Slovakian Roma forced to ghettos.' In: *The Christian Science Monitor* (January 3, 2003).

Frankfurter Allgemeine Zeitung (2004), 'Slowakische Regierung reagiert auf Roma-Unruhen' [Slovak government reacts to Roma riots] (March 1, 2004).

Fraser, A. (1992), *The Gypsies*. Oxford: Blackwell Publishers.

Government of the Slovak Republic (2003), *Štatút splnomocnenca vlády Slovenskej republiky pre rómske komunity* (vládou SR schválený 17. decembra 2003, uznesenie č. 1196/2003) [Status of the Slovak government plenipotentiary for the Roma community, decision Nr. Nr. 1196/2003], Paragraph 3, 1.

Gronemeyer, R. (ed.) (1983), *Eigensinn und Hilfe. Zigeuner in der Sozialpolitik heutiger Leistungsgesellschaften* [Stubborness and help. Gypsies in the social policy of today's meritocracy], Gießen.

Guy, W. (2001), 'The Czech lands and Slovakia: Another false down?' In: Guy, W. (ed.), *Between past and future: The Roma of Central and Eastern Europe*. Hatfield: University of Hertfordshire Press, pp. 285–323.

Holomek, K. (1998), 'Roma in the Former Czechoslovakia.' In: *Helsinki Citizens' Assembly – Roma Section Newsletter*, no. 4 (Juni 1998), 3f.

Hriczko, I., Magdolenova, K. (2003), 'Hazardous Games with Poverty.' In: *Romské Listy*, bimonthly supplement to 'Domino Forum' (19.12.2003).

Hurrle, J. (1998), 'On the outskirts of Košice.' *Newsletter of the HCA Roma Section*, No. 6, Brno: Helskinki Citizens' Assembly's Roma Section, pp. 12–15.

Ivanov, A. (2002), *Avoiding the Dependency Trap. The Roma in Central and Eastern Europe*. Bratislava: UNDP Regional Bureau for Europe and the Commonwealth of Independent States, 84.

Jakoubek, M. (2003), 'Romské osady – enklávy tradiční společnosti' [*Romské osady* – Enclaves of traditional society]. In: Jakoubek, M., Potuška, O. (2003) *Romské osady v kulturologické perspektivě* [*Romské osady* in the perspective of cultural sciences], Brno: Doplnek.

Jakoubek, M., Potuška, O. (2003), *Romské osady v kulturologické perspektivě* [*Romské osady* in the perspective of cultural sciences], Brno: Doplnek.

Jurová, A. (2002), 'Historický vývoj rómskych osád na Slovensku a problematika vlastníckych vzťahov k pôde.' ('*Nelegálne osady*'). [The historic development of Roma settlements in Slovakia and the landownership issue. ('Illegal *osady*')]. In: *Člověk a společnost*, Nr. 4, 2002. URL (14.5.2004): Internetový časopis SAV Košice, http://www.saske.sk/cas/4-2002/jurova.html.

Jurová, A. (2003), 'The Roma from 1945 until November 1998.' In: Vašečka et al.

2003, pp. 45–62.

Kollárova, Z. (1992), 'K vývoju rómskej society na Spiši do roku 1945' [On the development of the Roma in Spiš Region till 1945]. In: Mann 1992, pp. 61–72.

Lewis, O. (1966), 'The Culture of Poverty.' *Scientific American*, 217: 4, pp.19–25.

Mann, A. (1992), *Neznámia Romovia. Zo života a kultury Cigánov-Romov na Slovensku* [Unknown Roma. From the life of the Gypsies-Roma in Slovakia], Bratislava: Ister Science Press.

Mayerhofer, C. (1987), *Dorfzigeuner. Kultur und Geschichte der Burgenland-Roma von der Ersten Republik bis zur Gegenwart* [Village Gypsies. Culture and history of the Roma in Burgenland Region from the First Republic till the presence]. Wien: Picus Verlag.

MG-S-ROM 2000: Roma and Statistics. Strasbourg: Council of Europe.

Mušinka, A. (2003) 'Roma Housing.' In: Vašečka et al., pp. 371–391.

Nečas, Č. (1992), *Českoslovenští Romové v letech 1938 – 1945* [Czechoslovak Roma in the years 1938 – 1945], Olomouc.

Novák, K. (…) 'Romská osada – tradice versus regres' [Romská osada – Tradition versus decay]. In: Jakoubek, M.; Potuška, O. *Romské osady v kulturologické perspektivě* [Romské osady in the perspective of cultural sciences], Brno: Doplnek, pp. 31–40.

Radičová, I. (ed.) (2004), *Atlas rómskych komunit na Slovensku 2004* [Atlas of Roma Communities in Slovakia], Bratislava: SPACE, IVO, KcpRO.

Radičova, I. (2003), 'The Roma on the Verge of Transformation.' In: Vašečka et al., pp. 63–74.

Revenga, A., Ringold, D., Tracy, W. (2002), *Poverty and Ethnicity: A Cross Country Study of Roma Poverty in Europe.* Washington, DC: The World Bank.

Ringold, D., Orenstein, M.A., Wilkens, E. (2003) *Roma in an Expanding Europe. Breaking the Poverty Cycle.* Washington: World Bank.

Scheffel, D. (1999), 'The Untouchables of Svinia.' In: *Human Organization*, 58: 1, pp. 44–53.

Scheffel, D. (2004), 'Slovak Roma on the threshold of Europe,' In: *Anthropology Today*, 20: 1, pp. 6–12.

Sobotka, E. (2001), 'Crusts from the table: Policy Formation towards Roma in the Czech Republic and Slovakia.' In: *Roma Rights*. Budapest: European Roma Rights Center.

UNDP (2004), *Millenium Development Goals Report. Slovak Republic*. Bratislava: UNDP Regional Bureau for Europe and the Commonwealth of Independent States.

Vašečka, M., Jurásková, M., Nicholson, T. (2003), *Čačipen pal o Roma. A Global Report on Roma in Slovakia.* Bratislava: Institute for Public Affairs (IVO).

Vašečka, M., Jurásková, M., Kriglerová, E., Puliš, P., Rybová, J. (2002), *Evaluation of the activities of the office of the plenipotentiary of the government.* Bratislava: Institute for Public Affairs (IVO).

Vláda Slovenskej Republiky (1999), *Stratégia vlády SR na riešenie problémov rómskej národnostnej menšiny, Oblasť bývania* [Strategy of the Slovak government for the solution of the Roma minority's problems. Sector Housing]. URL (14.6.2004): http://www.vlada.gov.sk/INFOSERVIS/DOKUMENTY/ROMSTRAT/sk_rs_strat_ob5.shtml,

Vláda Slovenskej Republiky (2001), *Comprehensive Development Programme for Roma Settlements.* URL (12.5.2004): http://www.government.gov.sk/orgovanova/dokumenty/rozvojovy_program_romskych_osad_en.doc.

World Bank, Foundation SPACE, INEKO, Open Society Institute (2002), *Poverty and Welfare of Roma in the Slovak Republic.* Washington, DC: The World Bank

Zoon, I. (2001), *On the Margins. Roma and Public Services in Slovakia.* Budapest: Open Society Institute.

Chapter 10

Urban Development in Hungary After 1990

Zoltán Dövényi and Zoltán Kovács

Introduction

Prior to 1990 urban development in Hungary – similarly to other spheres of the socio-economic system – was influenced and controlled to a large extent by the state. Just like in other state socialist countries, urban development was a central focus of the planning system. After World War II industry and industrialisation were considered by the communist regimes in Central and Eastern Europe to be the only possible way to catch up with the west. Cities and industry became symbols of modernity while villages and agriculture meant the evil backward past (Grime and Kovács 2001). As a consequence emphasis was placed on the development of heavy industry from the early 1950s onwards which in turn resulted in continuous concentration of the population and very dynamic urban growth. Urban processes such as urban sprawl or gentrification of inner urban neighbourhoods that are characteristic for the western part of Europe were unknown here (Kovács 1999). The collapse of state socialism generated far-reaching social and economic transformations in Central and Eastern Europe after 1990. These processes led to fundamental changes in the urban system and the spatial organisation of cities. Thus 1990 represented a new era in the urban and regional development of these countries (Fassmann 1997).

The main aim of this chapter is to provide a broad overview of the spatial characteristics of urban development in Hungary. In the analysis special attention is paid to the latest trends in urbanisation, processes that began with the socio-economic transformation. We ask whether processes of urban development typical for the Western part of Europe are also present here. Is there a convergence in urbanisation between Hungry and the West or is divergence still dominant?

Since knowledge on the historical aspects of urban development is vitally important for a comprehensive understanding of current processes, we start this essay with a short overview of the historical background of urbanisation and urban development in Hungary.

Historical Background of Urban Development in Hungary

Apart from mediaeval and early capitalist urbanisation the modern urban network of Hungary came into being relatively late, only in the period of the Austro-Hungarian Monarchy (1867–1918). In this historical period the population in cities grew three times faster than the one in villages, a clear sign for the acceleration of urbanisation. The extraordinary growth of the capital Budapest was especially remarkable. A large part of the national industry, finance and trade, as well as science and culture were concentrated in the capital in a manner previously unprecedented in Central Europe. By 1910 Budapest had become the eighth largest city in Europe and the absolutely dominant centre of the country. As a consequence of the buoyant urban growth a wide zone of suburban settlements developed around the city. Although urban development in general was very dynamic in the country in this period, the dynamism of Budapest could not be matched by other centres. If we take indicators of economic growth and dynamism of the ten regional centres of Hungary at that time – i.e. Zágráb (Zagreb), Kolozsvár (Cluj), Pozsony (Bratislava), Szeged, Kassa (Košice), Debrecen, Pécs, Temesvár (Timisoara), Nagyvárad (Oradea), Arad – the aggregate figures of these cities still fall far behind the level of Budapest.

Prior to World War I one of the most striking characteristics of the settlement network in Hungary was the discrepancy between the numbers of legally defined 'towns' and those places that could be functionally considered as cities. In 1900 altogether 131 Hungarian settlements had the legal status of towns, but the number of central places that functionally met the criteria of cities was about 250 (Beluszky 1990). If we take the latest figure, roughly one third of the Hungarian population lived in urban places at the turn of the 20th century as opposed to the officially registered urban ratio of 13.8 per cent. If we look at the urban hierarchy of Hungary at the beginning of the 20th century we can conclude that the most typical feature of the urban system was – in addition to the dominant position of Budapest – the weakness of small towns. This feature remained practically intact until the political changes of 1989/90.

World War I and the subsequent Trianon Peace Treaty of 1920 altered the conditions of spatial development of the country. Due to the treaty Hungary lost 71 per cent of its territory and 64 per cent of its population. By virtue of the agreements fixed in the peace treaty a new regional order was established in the Carpathian Basin and the socio-economic character and urban pattern of Hungary were changed fundamentally. In 1918 there were 139 settlements with the status of towns in Hungary, and only 47 after 1920. Another important characteristic of the Hungarian urban network was the extreme increase of the weight of Budapest as a primate city. From one day to another Budapest became the capital city of a country of 7,6 million inhabitants – compared to nearly 21 million in what had been Hungary before 1920. In 1910 less than 5 per cent of the total Hungarian population lived in Budapest; in 1920 it was 12 per cent.

Due to the Trianon Peace Treaty Hungary lost seven out of the ten major regional centres, and only Szeged, Debrecen and Pécs remained within the limits of the new boundaries (Figure 10.1). As a consequence of the territorial changes Budapest became the absolutely dominant centre, or as many call her, the 'hydrocephalus' or 'swollen head' of the country. Beside Budapest no other major centres could develop, and typically the second largest towns (Debrecen and Szeged) had only about a tenth of the size of the capital.

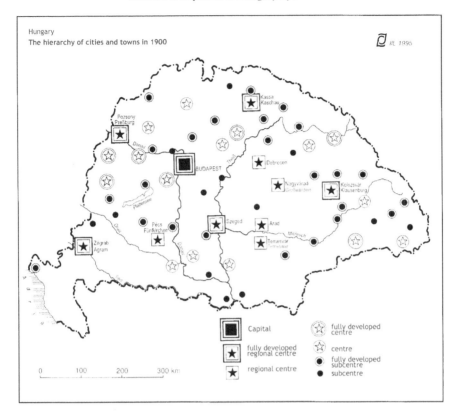

Figure 10.1   The hierarchy of Hungarian cities and towns around 1900
Source: Berényi and Dövenyi 1996

The urbanisation process of the inter-war period concentrated mostly on Budapest and its close surroundings. The development of other towns slowed down, not least because in many cases the new borderline divided urban centres from large parts of their hinterlands, cutting the organically grown connections. This, of course, had serious economic and social consequences as the development of these towns has been retarded significantly and the rate of growth of the population in these towns fell behind the national average. Thus, the ratio of urban population showed hardly any increase between 1920 and 1940 (31.8 per cent and 34.6 per cent respectively).

After the radical territorial reorganisation of the country the development of the Hungarian urban network did not change too much in the interwar period compared to the previous decades. There was no serious counter-pole to Budapest that could counterbalance the dominant role of the capital. The discrepancy between the legally and functionally defined towns remained intact. At the end of World War II 56 Hungarian settlements had the legal status of town, whereas the number of centres that functioned as towns reached about 150 (Beluszky 1973). The lack of a functionally well developed small-town network also remained a permanent problem.

When evaluating the urban development of Hungary in the interwar period, it should also be mentioned that the new national border did not respect the hierarchical evolution of the settlement network. Important regional centres such as Nagyvárad (Oradea), Kassa (Košice), Sopron, and Komárom lost various parts of their zones of influence, while large areas were left devoid of a significant urban centre. These processes also affected neighbouring countries – both sides of the Hungarian national boundary were hit hard by economic stagnation and depression (Berényi and Dövényi 1996).

The Impact of State Socialism on Urban Development in Hungary between 1950 and 1990

After World War II, a communist-type centrally planned economy and a single party system similar to other Central and Eastern European countries were introduced in Hungary. Urban and regional planning in the state socialist countries had – at least in the 1950s and 60s – two major goals, namely industrialisation and urbanisation. Industrialisation meant not only the fostered development of mining and production but also the internal re-organisation of agriculture, i.e. the establishment of large scale agricultural plants (state farms, cooperatives). Mainly large- and medium-sized towns were designated for industrial development, but there were also some new towns established specifically for industrial purposes. In the first long-term Hungarian urban development strategy published in 1962, cities were classified by planners according to their capacity for accommodating industry (Enyedi 1996). This meant that their development prospects were designated according to this single criterion. Through this kind of strictly controlled urbanisation the communist state also tried to fulfil its main societal goals, the continuous enlargement of the industrial working class and the abolishment of smallholders.

One of the most important characteristics of urbanisation and urban development in the period between the end of the 1940s and 1990 was the sharp increase in the number of towns. The systematic use of the 'legal factor' in urbanisation was partly connected with growing state intervention and the centrally planned character of modernisation initiated from above. Between 1950 and 1965 the promotion of villages to towns was quite limited and only 9 settlements were designated towns. The majority of them were so-called socialist industrial towns, settlements developed most often around an industrial estate or mine. Typical examples are Dunaújváros (former Stalin-city, with a large steel works on the Danube south of Budapest), or Tiszaújváros (former Lenin-city with petrochemical industry), Oroszlány (coal-mining), Komló (coal-mining) etc. Another important element of the urban development strategy in the 1950s and 60s was the modernisation and intense development of old industrial centres, e.g. Ózd, Tatabánya, Salgótarján.

Following the first comprehensive urban and regional development strategies of the 1960s, more industry was then located in provincial towns, accelerating the development of the whole urban system. The National Settlement Development Plan (OTK) approved in 1971 specified a strict sequence of order among settlements, including towns. As a consequence of the national development policy, in the following two decades the urban network was extended significantly. The majority of the newly designated towns had long traditions of urban functions and disposed of an excessive zone of influence. Therefore the allocation of town status was rather an adjustment of the administrative structure of the country to the organic development of the settlement system than a mere legal step. As a consequence, one of the major contradictions of the Hungarian urban system deriving from the difference between the number of towns in legal and in functional sense gradually disappeared. By the second half of the 1980s the Hungarian urban system already comprised most of those settlements that were in reality functioning as cities (Figure 10.2).

Due to the increase in the number of towns the level of urbanisation has also grown. In 1949 37 per cent of the total population lived in the then 54 cities, thus, Hungary was still predominantly a rural country compared to the West. After that the urban ratio of the country increased steadily and reached 53 per cent by the time of the 1980 census. In the 1980s the proportion of the urban population grew further and 62 per cent of the population lived in cities by 1990.

The growth of the urban population and the intensity of rural-urban migration were fairly uneven during the state socialist period. Rural-urban migration peaked in the 1950s and 1960s when over a million people left the villages and moved to cities as a consequence of the forced collectivisation of agricultural land and the extensive development of heavy industry. As a result, the population of Budapest grew by 200,000, and the one in other cities by 700,000 in the 1960s. However, rural-urban migration slowed down gradually from the early 1970s, and between 1970 and 1980 there were already some cities (most of them located on the Great Hungarian Plain) where the balance of migration turned negative. In the 1980s the attractivity of cities declined further and between 1980 and 1990 the cities and towns already lost population – a totally new phenomenon which modern Hungary had never experienced before except during the two world wars.

The net-population decrease in Hungarian cities can be explained basically by two factors. The general aging process and the subsequent decrease of the population (Hungary has had a natural decrease since 1981) hit cities more seriously, but the accelerating outmigration of younger and better educated groups also played an important role. From the mid-1980s, the attractivity of towns as a place of residence declined enormously (this is especially true for Budapest and other large cities) mainly due to economic recession and increasing costs of living. This resulted in a general tendency of outmigration from urban places.

The urbanisation process during the communist period was greatly influenced by the state. In the framework of the centrally planned economy cities played a distinguished role, the level of technical and social infrastructure provision was much higher than in the villages. The National Settlement Development Plan

(OTK) of 1971 introduced a strongly centralised model of settlement development and planning. This comprehensive development concept, specifying a strict functional hierarchy, was applied to the entire country and resulted in the creation of a hierarchical settlement network. It also constituted the framework within which development funds were distributed. Due to the concept more than 80 per cent of the financial resources were allocated to 150 centres (mainly cities) whereas the rest of the settlement system received hardly any development funds. Due to the growing dissatisfaction with the concept and its consequences for the municipalities the National Settlement Development Plan (OTK) was first modified and at last abolished in 1984. Moreover, in 1985 an administrative reform was introduced in Hungary, showing clear signs of decentralisation.

The privileging of cities under the communist regime led to two major processes within the settlement system: firstly, the difference between cities and villages increased, secondly, the difference within the urban network slowly decreased (homogenisation). Despite the considerable urban development after World War II large parts of the Hungarian countryside remained without an urban centre and were cut off from modern urbanisation. The frequency and size of such areas clearly increased from the West to the East which reflects the regional variation of urbanisation in Hungary. The urban network west of the Danube river is spatially balanced and well-established, whereas in Eastern Hungary, more specifically on the Great Hungarian Plain, it is less developed.

Towards the end of the state socialist period the political, economic and social life of Hungarian cities and their internal structure exhibited the following features which could also be found in other East Central European countries (Häußermann 1996, Standl 1998).

- Local decision making was fragmented between the party, the central state and industry, and there was a complete absence of local self-government. Cities were ruled by hand-picked councils which followed the instructions of the communist party.
- Despite the beginnings of de-industrialisation which were also supported by the central state, the manufacturing base of the cities remained strong, and the importance of the service sector was far below western standards.
- During state socialism urban land was transferred to state ownership, or at least largely withdrawn from private rights of disposal. Given the absence of free property markets land rents had no significance in urban development. As a consequence, socialist cities remained fairly compact with large, relatively homogeneous functional areas. But no western-style CBDs (Central Business Districts) with a strong service sector existed (Kovács and Wiessner 1999).
- The role of the state in the field of housing construction and renovation was absolutely dominant in the cities. New housing construction took place nearly exclusively at the outer fringe of the cities, mostly in the form of large housing estates. This high concentration of housing construction also enabled cities to remain compact. On the other hand the housing stock of the inner quarters

built before World War II deteriorated visibly, mainly due to neglect (Sailer 2001).

- In spite of the growing social differences, especially from the late 1960s onwards, the level of residential segregation remained relatively low. State intervention both in the labour and housing markets was rather strong; the main goal of social policy was homogenisation. Secure employment and cheap housing constituted the corner stones of the communist system, and they were thought to be the main tools for realizing the dream of a classless society.

Conditions of Urban Development after 1990

After 1989 due to the radical political and economic changes the conditions of urban development also changed fundamentally. The basis for the transformation of urban development was created by a far-reaching re-organisation that took place in the political, economic, social and planning systems (Kovács 1998).

As far as the *political transformation* is concerned the dissolution of the Warsaw Pact, the withdrawal of the Soviet army and the subsequent re-establishment of political sovereignty played a decisive role. Thanks to this newly-found political sovereignty the isolation of Hungary and other East Central European countries came to an end and cities once trapped in their national spaces were now able to reintegrate themselves into the European and global urban systems. Alterations in the external geopolitical climate have also made far-reaching internal political changes possible. Among others, the revival of multi-party systems, free parliamentary elections held in March 1990 and the ousting of the communist party from power represented the major steps of political transition. With respect to urban change an important component of the political transformation was the return to self-governance and the subsequent shift of control from the central (state) to the local (community) level. Act 65/1990 ('Act on Local Governments') lifted the legal discrimination of villages. On the other hand, the Act granted settlements more power to control and influence their own development.

The *economic transformation* that meant a radical shift from central planning to a free-market system had also several components. First the collapse of the COMECON proved to be a kind of 'shock therapy', leading to bankruptcy and mass-liquidation of many companies. The fall of the 'Iron Curtain' symbolised not only the advent of liberty, but gave the global economy and its main actors, the transnational corporations, direct access to the Hungarian market. The appearance of western firms bringing foreign capital investment and modern technology to the region constituted an important driving force in the economic restructuring of cities and regions (Turnock 1997). Besides the infiltration of global capital, the disintegration and privatisation of large state companies – especially in the socialist heavy industry – played an important role (by 1996 the private sector produced 70 per cent of the national GDP). The process of de-industrialisation and the expansion of the service sector speeded up considerably. Economic restructuring was very uneven geographically. Some regions and their cities (like Miskolc or Ózd), and especially the socialist industrial cities were hit hard by high levels of unemployment, growing poverty and mass-outmigration.

The *societal conditions* of urban development also changed considerably. Hungary has experienced a severe population decrease since 1980. The main reason has been natural decrease due to very low fertility rates exacerbated by the dramatic ageing of the society. In 1990 the annual number of births per one thousand inhabitants still amounted to 12. This figure decreased to 9.4 by 1999, then there was a gradual increase and by 2001 the level of births again reached 10 per thousand. The number of children per family is 1.3 which is astonishingly low. Due to unfavourable demographic conditions the population of the country decreased by 177,000 or 1.7 per cent between 1990 and 2001. The demographic conditions of the society show significant regional variations. A natural decrease of the population is persistent in Budapest and Western Hungary, whereas some eastern regions still show a modest natural increase. In terms of the internal structure of the society thanks to the growing differentiation of incomes, social differences in Hungarian cities also increased very rapidly. The remnants of social housing became more and more the shelter of the urban poor, whereas the better-off and the young started to leave the cities in great numbers and invaded the green suburbs, copying the suburbanisation processes of western cities in the 1960s and 1970s.

With regard to the *planning system* changes in the political and economic macrospheres led to the complete abolition of central planning. The mechanisms of central planning and the role of the National Planning Office (OT) were replaced by the regulatory mechanism of the market. Market forces, demand and supply became the main factors organising the production and exchange of goods. This shift brought about immediate improvements in the supply of services in rural areas and other less privileged towns. Through the democratic transition most of the planning competencies were delegated to the local level (i.e. cities, in Budapest districts). The shift of control from the central to the local level also meant that cities could adopt their own development policies, adjusted to their potentials and priorities. The whole concept of regional development and planning lost its earlier influence, and it gained momentum again only after 1996 when the new 'Act on Regional Development' was passed.

Main Features of Urban Development after 1990

After 1990 the political and economic transformations induced significant changes in the urban system as well. Due to a dramatic increase of regional disparities within the country, differences within the urban network also started to grow. Together, all these developments led to new forms and phenomena of urbanisation. In the following paragraphs we sum up the most important features of urban development in Hungary since 1990.

Extension of the Urban System

As part of the democratisation process the legal promotion of villages to the status of town became much simpler. In Hungary an index of urbanisation is applied for the legal definition of towns. This is a mix of 175 different indicators taking into account the functional role which the settlement plays in the settlement network, its historical traditions as well as indicators that reflect the level of infrastructural development (e.g. the proportion of paved roads or dwellings with bathrooms). Communes that want to reach town status need to submit an application to the Ministry of the Interior. Applications are evaluated by a special expert commission set up by the Ministry. The president of the state will grant urban status to those settlements that meet the majority of the criteria.

The size of population is only one of the 175 indicators; therefore it plays a rather limited role in the legal definition of towns. The smallest town of the Hungarian urban network is Zalakaros with 1,566 inhabitants, which is a holiday resort with a thermal spa in Western Hungary. Due to the liberal system of legal definition of towns 108 settlements have been granted town status in Hungary since 1990. Consequently, the number of urban places – at least in legal sense – has grown to 274 in Hungary by now.

The extension of the urban system shows interesting variations over time. According to the law, in the years when parliamentary elections are taking place (every 4 years since 1990) no new towns can be created. However, if we analyse the interval periods between two elections we can clearly observe increasing activities with respect to the promotion of villages to towns. The number of newly created towns regularly culminates in the last year before the elections (the first exception seems to be 2004, when on the 1st of July 18 Hungarian villages were granted town status in legal sense at once). The reason is clear, governmental parties want to reward settlements with loyal mayors or local governments consisting of the same parties. There are strong lobby mechanisms in each parliamentary party (right now four) independently from their ideological platforms that support this mechanism to strengthen the party's political base. Since the probability of re-election is very low (all four post-communist elections resulted in new governments), governments use the opportunity to upgrade those municipalities which support their party. This can even yield additional votes in the new election.

Why are municipalities so motivated to become towns, once urban places are not privileged by politics any more, and villages get practically the same amount of per capita funds for communal and infrastructural development as cities? Empirical research revealed that settlements that attain town status are more successful in getting funds from development programmes (e.g. the development of a gas network or a sewage system) than those of the same size that do not have this status. Since development funds are distributed predominantly through application procedures it means that municipalities with town status have a better chance to apply successfully. This is basically a consequence of the local government system. Towns are obliged to maintain special departments responsible for urban development and planning, and according to the law they must also set up their own master plan, which is an important criterion for any application. Villages do not have to set up a master plan and therefore do not have to bear the respective costs. Investments in local planning capacities seem to yield advantages for the settlements.

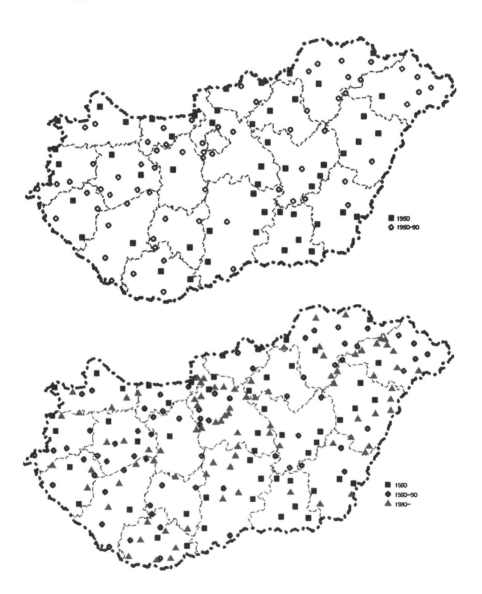

Figure 10.2a–b   Cities and towns in Hungary, 1990 and 2004
Source: Z. Kovács and L. Kaiser

Transformation of the Urban System

Thanks to the intensive legal upgrading of villages to towns the Hungarian urban system became spatially more dense and balanced after 1990. In 1990 there was one town in every 560 km² of the country, in 2004 already in 340 km². Vast areas without towns practically do not exist any more in Hungary, we can even distinguish regions with astonishingly high densities of cities (e.g. the urban region of Budapest, or the region around Lake Balaton) (Figure 10.2).

After 1990 the composition of the urban system has also changed considerably. In 1990 there were only 43 towns in Hungary with less than 10,000 inhabitants. Within this group the segment of the smallest towns (the so-called 'dwarf towns', i.e. towns with less than 5,000 inhabitants) was rather limited, with only 6 such settlements. The biggest group of towns was the group of so-called 'large-small towns' (i.e. towns between 10 and 20 thousand inhabitants) with altogether 62 members.

The large-scale extension of the urban system in the last 15 years has resulted in the growing number of small towns (i.e. towns below 10,000 inhabitants). At the beginning of 2004 there were already 120 towns that belonged to this category in Hungary. Within this size category we have been able to observe the mushrooming of 'dwarf towns'. In 2004 there were already 40 municipalities with town status in Hungary with less than 5,000 inhabitants.

The spatial distribution of small and dwarf towns in Hungary is rather polarised. East of the Danube river, in the Great Hungarian Plain, the main type of cities are the so-called agrarian towns with a relatively high number of inhabitants (above 10,000). On the other hand in Northern and Western Hungary market towns with strong central functions and long urban traditions are typical. Among these cities small and dwarf towns are clearly overrepresented.

The large-scale enlargement of the lower levels of the urban hierarchy resulted in a changing distribution of the urban population. In 2004 more than one quarter (26,6 per cent) of Hungarian urban dwellers lived in towns with less than 20,000 inhabitants. In 1990 this ratio had amounted to only 18,8 per cent. The extraordinary growth of the share of towns with less than 10,000 inhabitants is even more striking. In 1990 only 4,8 per cent of the Hungarian urban population lived in such small towns, in 2004 already 11 per cent.

Summing up, the political liberalisation and democratisation in Hungary led to a dynamic increase in the number of towns. On the other hand this process brought about a continuous devaluation of the town status. A significant part of the settlements that were granted town status after 1990 functionally and infrastructurally can not be qualified as real towns, most of them remained villages in many respects.

Changing Levels of Urbanisation

After long decades of constant urban growth and rising urban ratios perhaps the most striking phenomenon in Hungary is the relative decline of the urban population. In 1990 62 per cent of the Hungarian population lived in cities. The national urban ratio increased to 64 per cent by 2004, however the number of towns also grew from 166 to 256 during the same period. If we take the total population of the 166 towns that attained urban status by 1990, the level of urbanisation would have been only 58.6 per cent in 2004.

The relative decline of the urban population can be explained basically by two main factors. On the one hand, cities (and especially Budapest) were hit by a natural decrease of the population much harder than villages. In cities birth rates are generally lower and death rates are higher than in villages, which is especially true for the bigger cities and the capital.

On the other hand, large cities with increasing pollution and congestion, and decreasing security have become rather unpopular places to live in. Due to decreasing satisfaction there has been a growing out-migration from urban areas after 1990 (fig. 10.3). However, the reasons for outmigration differed significantly by regions. In cities in regions with strong industrial traditions (i.e. Northern Hungary and north of Lake Balaton) the collapse of the socialist industry and the concomitant economic crisis generated mass-outmigration to rural areas, especially among unskilled and retired people (e. g. Miskolc Ózd, Salgótarján). Meanwhile, there is also outmigration from the city centres in to the suburbs in economically more prosperous cities, analogous to the suburbanisation processes of west European cities (Burdack, Dövényi and Kovács 2004; Dövényi, Kok and Kovács 1997). Suburbanisation is especially pronounced around Budapest generating massive population losses for the capital city, highlighted above (Izsák and Probáld 2001).

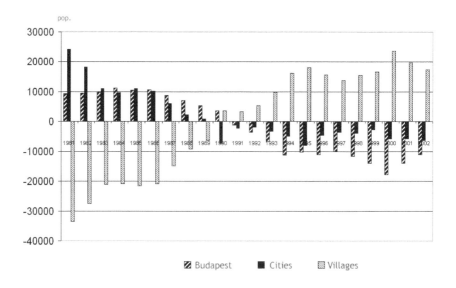

Figure 10.3   Net migration in Hungary by settlement types (1980–2002)

Due to a natural decrease and growing outmigration Budapest and the towns in the countryside lost 120,000 and 60,000 inhabitants between 1990 and 2001 respectively. If we look at the level of population loss within the settlement hierarchy the two settlement types most seriously touched by population decrease were Budapest and the small villages (i.e. settlements with less than 200 inhabitants). Demographically the most dynamic segment of the Hungarian settlement system is the group of municipalities between 5,000 and 10,000 inhabitants (Table 10.1). This group is dominated by small towns with strong tourist functions (e.g. places around Lake Balaton), suburbs around Budapest and other major cities (e.g. Miskolc, Debrecen, Győr), and small towns with dynamic business functions in North Western Hungary (e.g. Szentgotthárd – General Motors).

With stagnating urban ratios despite growing numbers of cities Hungary entered a new phase of urbanisation after 1990. The gap between towns and villages has generally decreased. This can be simply explained by the presence of market circumstances and the re-emergence of local governance. Before 1990, as in other centrally planned economies the distribution of resources for municipal development was determined by the central political and planning authorities. After 1990, the allocation of resources became more just, the privileges of towns were terminated, and villages gained more opportunities for infrastructural development. This simply resulted in improving local infrastructure, and better services and living conditions in rural areas, enabling villages to keep their inhabitants and attract new ones (Meusburger 2001).

Table 10.1   Changes of population in different settlement categories in Hungary (1990–2001)

Settlement size	1990	2001	%
Budapest	2,016,681	1,777,921	88.2
above 100,000	1,203,603	1,188,232	98.7
50,000 – 99,999	776,628	771,679	99.4
20,000 – 49,999	1,216,819	1,208,284	99.3
10,000 – 19,999	1,079,331	1,098,962	101.8
5,000 – 9,999	893,739	932,443	104.3
2,000 – 4,999	1,451,860	1,505,513	103.7
1,000 – 1,999	925,580	939,724	101.5
500 – 999	511,442	497,655	97.3
200 – 499	254,689	239,934	94.2
below 200	44,451	37,968	85.4
Total	**10,374,823**	**10,198,315**	**98.3**

Source: Census 1990, 2001

Differentiation within the Urban System and Restructuring of Cities

Due to the transformation new types of differentiation and polarisation emerged within the urban system. The differentiation process of the urban network very much depended on geographical location, local conditions like the economic structure or the level of human capital, and the talent of local politicians (Kovács and Dövényi 1998).

Apart from Budapest, cities which have been able to compete successfully for foreign direct investment (automobile industry, electronic equipments etc.) and for the location of international businesses and cultural functions are located nearly exclusively in Western Hungary. The best examples are Győr and Székesfehérvár (each having attracted over 1 billion US$ of foreign direct investment since 1990), but Szombathely, Veszprém, Kőszeg or Sopron could also be mentioned. These cities together with Budapest enjoy 'gateway' functions within the country and belong to the clear winners of the transformation (Rudolph, Hardi and Terpitz 2002). This group of cities provides the growth axis of the new economic regime in Hungary.

Cities lying outside this belt have not been able to take advantage of the proximity of western markets and this is especially obvious if we take the stagnation or relative decline of the cities lying in the eastern border belt. The losers of the post-socialist transformation are the group of socialist cities (e.g. Oroszlány, Komló) and cities heavily industrialised during the years of communism (e.g. Miskolc, Salgótarján), as well as the agricultural towns of the Great Hungarian Plain (e.g. Hajdúböszörmény, Mátészalka, Kisvárda) with a lack of investment and a high rate of unemployment (Kovács and Dövényi 1998).

Economic restructuring and increasing polarisation within the society have also generated transformations within cities. The introduction of a capitalist economy meant that the market rather than government planning became the principal allocator of land and money. As a consequence the landscape and the internal structure of Hungarian cities have undergone tremendous changes since 1990. There are two ecological zones where we can observe high dynamism in urban development: the inner city and the periphery (Kovács and Wiessner 1997).

The economic restructuring has led to a growing demand for non-residential (business, office etc.) space especially in inner-city locations. The re-establishment of the real estate market, based upon land rents, has led to a rapid functional conversion, from residential to business use in the centre of Hungarian cities (Kovács 1994). The post-socialist urban development has been overwhelmed by rapid and intensive reinvestment at the urban core, which resulted in the massive expansion of the city (Kovács and Wiessner 2004). There is an obvious connection between the functional change and revitalisation of inner city neighbourhoods and the growing integration of these places into the world economy.

A deregulation of the housing market and the growth of income differentials are producing new forms of polarisation of urban spaces. Social housing especially in the form of high rise housing estates and old working class tenement blocks is becoming more and more the shelter of the urban poor (Wiessner 1997). The mixture of different social groups so characteristic of much of the fabric of the Hungarian cities has been disappearing fast. As part of the social differentiation we can observe the rise of the neighbourhoods of the 'newly-rich' sometimes close to the city centre where the old and less affluent populations are being displaced. But the better-off also move outside the physical confines of the city where development control policies are more relaxed. The new suburbs with their luxurious environment stand in sharp contrast with the decaying inner urban neighbourhoods or high-rise housing estates (Kovács and Wiessner 1999).

The dramatic increase of car ownership generated a high intensity of short distance migration. Suburbanisation and suburban commuting are well-known spatial phenomena in post-communist Hungary and present around all major cities and even smaller towns (e.g. Százhalombatta, Szentendre, cf. Berényi 1997, Kok & Kovács 1999, Timár & Váradi 2001). The relative de-concentration of the population resulted in new spatial patterns of daily commuting within the country.

A specific feature of suburbanisation in Hungary is that not only middle-class families, but also lower class and elderly people are leaving the cities, as they are suffering from rising costs of living they can hardly afford (Dövényi, Kok and Kovács 1997). In addition to residential suburbanisation we can distinguish clear signs of relocation of service functions to the urban periphery. The newly erected shopping and leisure centres as well as office locations are mostly realised in the form of green field investments. This process resulted in rapidly expanding suburbs, forming new growth poles at the periphery of cities. These new multi-functional business complexes constitute a great challenge for the traditional inner cities (Burdack, Dövényi and Kovács 2004).

Conclusions

The collapse of communism and the subsequent changes in the political and economic system have resulted in a radical transformation both of the Hungarian urban system and within the individual cities. The year 1990 signals a new phase of urbanisation in Hungary and other East Central European countries.

One of the bases of the communist system was the highly centralised distribution of resources through the channels of central planning. This resulted in large-scale homogenisation not only within society but also within the urban system. Thanks to this model of urban and regional development the landscape of cities also became very similar, and the result was what many called the 'socialist city'. The introduction of a market economy induced rapid differentiation processes both within the urban system and inside the individual cities.

Finally, the question should be answered as to how the ongoing processes of urban development in Hungary relate to global trends of urbanisation. If we carefully analyse the Hungarian experience of urbanisation after 1990 it can be said that there is a considerable delay in the form and intensity of spatial processes (i.e. concentration versus deconcentration) compared to Western Europe. Suburbanisation, which was so typical in the west after the post World War II economic boom, reached Hungary only at the end of the 1980s. However, by the mid-1990s clear trends of desurbanisation (i.e. the rapid growth of remote rural areas at the expense of traditional urban regions) emerged. Moreover, by the late 1990s (at least in Budapest and other major cities) the first signs of re-urbanisation became visible in the form of massive neighbourhood gentrification. Thus, cycles of the urbanisation model are compressed in time, and they are present in the spatial development of the country simultaneously.

References

Beluszky, P. (1973), 'Adalékok a magyar településhierarchia változásához, 1900–1970' (Contributions to the changes of Hungarian settlement hierarchy, 1900–1970), *Földrajzi Értesítő* 21, pp. 121–142.

Beluszky, P. (1990), 'Magyarország városhálózata 1900-ban' (The urban network of Hungary in 1900). In: Tóth, J. (ed.) *Tér, Idő, Társadalom*. Pécs, pp. 92–133.

Berényi, I. (1997), 'Auswirkungen der Suburbanisierung auf die Stadtentwicklung von Budapest.' In: Kovács, Z. and Wiessner, R. (eds), *Prozesse und Perspektiven der Stadtentwicklung in Ostmitteleuropa*. Münchener Geographische Hefte 76, pp. 259–268.

Berényi, I. and Dövényi, Z. (1996), 'Historische und aktuelle Entwicklungen des ungarischen Siedlungsnetzes.' In: Mayr, A. and Grimm, F.D. (eds), *Städte und Städtesysteme in Mittel- und Südosteuropa*, Beiträge zur Regionalen Geographie 39. Leipzig, pp. 104–171.

Burdack, J., Dövényi, Z. and Kovács, Z. (2004), 'Am Rand von Budapest – Die Metropolitane Peripherie zwischen nachholender Entwicklung und eigenem Weg', *Petermanns Geographische Mitteilungen* 148, 3, pp. 30–39

Dövényi, Z.,, Kok, H. and Kovács, Z. (1997), 'A szuburbanizáció, a lokális társadalom és a helyi önkormányzati politika összefüggései a budapesti agglomeráció településeiben' (The connections between suburbanisation, local society and communal politics in the settlements of the Budapest Agglomeration). In: Illés, S. and Tóth, P.P. (eds), *Migráció* (Migration). Volume I. KSH Népességtudományi kutató Intézet. Budapest, pp. 229–237.

Enyedi, Gy. (1996), 'Urbanization under Socialism.' In: Andrusz, G., Harloe, M. and Szelenyi, I. (eds), *Cities after Socialism. Urban and Regional Change and Conflict in Post-Socialist Societies*. Oxford: Blackwell, pp. 100–118.

Fassmann, H. (1997), 'Veränderung des Städtesystems in Ostmitteleuropa.' In: Kovács, Z. and Wiessner, R. (eds), *Prozesse und Perspektiven der Stadtentwicklung in Ostmitteleuropa*, Münchener Geographische Hefte 76, pp. 49–61.

Grime, K. and Kovács, Z. (2001), 'Changing urban landscapes in East Central Europe.' In: Turnock, D. (ed.), *East Central Europe and the Former Soviet Union. Environment and Society*. London: Arnold, pp. 130–139.

Häußermann, H. (1996), 'From the Socialist to the Capitalist City.' In: Andrusz, G. et al. (eds), *Cities after Socialism*. Oxford: Blackwell, pp. 214–231.

Izsák, É. and Probáld, F. (2001), 'Recent differentiation processes in Budapest's suburban belt.' In: Meusburger, P. and Jöns, H. (eds), *Transformations in Hungary. Essays in Economy and Society*. Heidelberg: Physica Verlag, pp. 291–315.

Kovács, Z. (1994), 'A city at the cross-roads: social and economic transformation in Budapest.' *Urban Studies*, Vol. 31, No. 7, pp. 1081–1096.

Kovács, Z. (1999), 'Cities from state socialism to global capitalism: an introduction.' *GeoJournal*. Vol. 49, No. 1, pp. 1–6.

Kovács, Z. (2004), 'Socio-economic transition and regional differentiation in Hungary.' *Geographical Bulletin*. Tom. LIII. No. 1–2, pp. 33–49.

Kovács, Z. and Dövényi, Z. (1998), 'Urbanisation and Urban Development in Hungary.' In: Bassa, L. and Kertész, Á. (eds), *Windows on Hungarian Geography*. Studies in Geography in Hungary 29, pp. 157–173.

Kovács, Z. and Wiessner, R. (eds) (1997), *Prozesse und Perspektiven der Stadtentwicklung in Ostmitteleuropa*. Münchener Geographische Hefte 76.

Kovács, Z. and Wiessner, R. (2004), 'Budapest – Restructuring a European Metropolis.' *Europa Regional*. Vol. 12, No. 4, pp. 22–31.

Meusburger, P. (2001), 'Spatial and Social Disparities of Employment and Income in Hungary in the 1990s.' In: Meusburger, P. and Jöns, H. (eds), *Transformations in Hungary. Essays in Economy and Society*. Heidelberg: Physica Verlag, pp. 173–206.

Rudolph, R., Hardi, T. and Terpitz, A. (2002), 'Regionaltendenzen in Westungarn – Gewinner der Transformation? Győr und Pécs in den 1990er Jahren.' *Europa Regional*. Vol. 10, No. 4, pp. 154–165.

Sailer, U. (2001), 'Residential mobility during transformation: Hungarian cities in the 1990s.' In: Meusburger, P. and Jöns, H. (eds), *Transformations in Hungary. Essays in Economy and Society*, Heidelberg: Physica Verlag, pp. 329–354.

Standl, H. (1998), 'Der post-sozialistische Transformationsprozess im großstädtischen Einzelhandel Ostmittel- und Osteuropas.' *Europa Regional* Vol. 6, No. 3, pp. 2–15.

Timár, J. and Váradi, M.M. (2001), 'The uneven development of suburbanization during transition in Hungary', *European Urban and Regional Studies*. Vol. 8, No. 4, pp. 349–360.

Turnock, D. (1997), 'Urban and regional restructuring in Eastern Europe: the role of foreign investment.' *GeoJournal* Vol. 42, pp. 457–464.

Chapter 11

Cyprus, EU Accession and the Development of Tourism

Richard Sharpley

Introduction

Strategically located in the eastern Mediterranean, Cyprus has frequently been described as lying at the crossroads of three continents. As a result, the island has, throughout its history, succumbed to a succession of foreign powers. The Egyptians, Persians, Romans, Arabs, Venetians, Ottomans and, latterly, the British were all former rulers of Cyprus, each contributing to the country's now rich and diverse cultural heritage. However, since 1960 when, for the first time, the island became independent, it has evolved rapidly into an economically and socially developed nation. In terms of per capita GDP, for example, it is now the third wealthiest Mediterranean country after France and Italy (PIO 2001) whilst, according to other typical development indicators, such as access to health care, life expectancy, infant mortality rates, literacy, and educational attainment, Cyprus is classed amongst those countries which enjoy 'high human development'. Indeed, in 1999 the island ranked 25th on the UNDP's Human Development Index, above countries such as Portugal, Singapore and Malta (Sharpley 2003). More significantly, perhaps, with accession to the EU in May 2004, Cyprus has finally escaped its past by joining the European family as an equal to the European powers that once dominated the region.

Underpinning this remarkable socio-economic transformation in Cyprus has been the successful and no-less remarkable development of tourism. In 1960, annual tourist arrivals totalled just 25,700; over the next 13 years, arrivals grew by an annual average of over 20 per cent, reaching 264,000 in 1973, and tourist receipts grew by an annual average of 22 per cent, thereby providing the basis for rapid economic growth and development in Cyprus (Kammas 1993). The Turkish invasion and subsequent occupation of the northern sector of Cyprus in 1974 had a devastating impact on the island's economy in general and on the tourism sector in particular (Andronikou 1979; Gillmor 1989; Lockhart 1993) but, nevertheless, the tourism industry was rebuilt rapidly and successfully. By 1994, over 2 million tourists were visiting the island each year and, in 2001, a record 2.7 million arrivals were recorded. In that year, receipts from tourism amounted to CY£1,272 million, representing almost 22 per cent of total GDP. Thus, since independence in 1960, tourism has been, and continues to be, the catalyst for economic and social development on the island.

The development of tourism in Cyprus is explored elsewhere in detail (for example, Ayers 2000; Cope 2000; Ioannides and Holcomb 2001; Sharpley 2001a,b; 2003). However, the important point here is that, despite its fundamental role in driving the nation's socio-economic development, tourism on the island faces an uncertain future with a current decline in tourist arrivals and receipts threatening future sustainable economic growth. In 2003, for example, tourist arrivals were about 14 per cent down and receipts over 20 per cent down on the 2001 figures (Republic of Cyprus 2004). This decline in tourism can, in turn, be related to the manner in which tourism development has been planned and managed on the island over the last three decades, particularly with respect to coastal land-use planning and infrastructure development. Nevertheless, EU accession is seen by some as offering significant opportunities for the revitalisation of the tourism sector (Blake et al. 2003) although such optimism may be misplaced. The purpose of this chapter, therefore, is to explore both the manner in which tourism development in Cyprus has been planned and managed and, within that context, the implications of recent EU accession for the future health of the tourism sector. Firstly, however, it is useful to review briefly the political-economy of Cyprus in general and the development of tourism in particular as a framework for considering tourism planning and policy processes and the future development of the tourism sector.

Cyprus: Economy, Government and Planning – An Overview

The Economy and Tourism

At the time of independence in 1960, the new government inherited an economy that displayed the structural weaknesses of under-development. Agriculture was the dominant economic activity, providing almost half of all employment on the island but contributing just 16 per cent of GDP. At the same time, manufacturing was restricted to the processing of locally produced agricultural raw materials, contributing about 10 per cent of GDP. Minerals constituted over half of all exports, yet the mining industry was in decline as a result of resource depletion, and agricultural products constituted a further 32 per cent of exports. Tourism, meanwhile, was in its infancy (Andronikou 1987). Importantly, hidden unemployment and underemployment were widespread and the country was suffering a significant outflow of labour and financial capital (PIO 2001: 149). Therefore, 'one of the major goals of the island's first development program called for an import substitution policy as a tool to achieve industrialisation and greater economic development...taking advantage of the island's natural resources' (Kammas 1993: 71).

This was certainly achieved. As noted above, the economy of Cyprus has developed remarkably; 'in the 14 years after independence in 1960, the Republic of Cyprus, with a free enterprise economy based on trade and agriculture, achieved a higher standard of living than any of its neighbours, with the exception of Israel' (Brey 1995: 92). By 1995, per capita GDP had reached US$11,540, slightly higher than Greece (US$10,566) and significantly higher than Turkey (US$2,853). Since then it has continued to grow, reaching an estimated US$17,847 in 2003 placing the country 16th in the world order.

The relative contribution of different sectors of the economy has also undergone a fundamental transformation. Agriculture accounts for a continually declining proportion of GDP (currently less than 5 per cent) and about 9 per cent of employment, though agricultural products (raw and processed) still account for some 30 per cent of exports. Conversely, construction, manufacturing and the tertiary sectors have come to occupy an increasingly important position in the economy with tourism, in particular, dominating economic activity. In a country that enjoys virtually full employment, the tourism sector accounts for 25 per cent of all employment whilst direct tourism expenditure contributes around 20 per cent of GDP and 40 per cent of exports. However, as Ayers (2000) points out, these figures do not indicate the true impact of tourism on the economy. That is, the rapid growth in tourism has stimulated growth in other sectors, particularly construction, as well as in related industries such as financial services, communications, and transport, whilst 'agriculture and manufacturing also benefited from the increasing number of arrivals who boosted demand for a wide range of locally produced products' (Ayers 2000). At the same time, the production of other products and handicrafts, such as wines and lace, has been revitalised by tourism demand. As a result, according to the World Travel & Tourism Council's satellite accounting research, the tourism economy in Cyprus contributed almost 31 per cent of GDP in 2001 (WTTC 2001).

There can be no doubting, therefore, the contribution of tourism to the Cypriot economy, although it has long been recognised that the island's dependence on tourism and, more generally, the character of tourism development in Cyprus is unsustainable in the longer term (EIU 1992). The challenges facing the tourism sector and current plans, reflecting European policies, to achieve more sustainable tourism development are discussed shortly. However, the following section reviews broader governmental structures and planning processes within which tourism planning and policy-making occurs and which has, implicitly, 'permitted' tourism development to follow an unsustainable path.

Government and Planning

The political system in Cyprus has been described by one commentator as 'excessive democracy' (Cousaris 1998). More specifically, the Republic of Cyprus is an independent sovereign state with a presidential system of government. The President, elected to a five-year term, exercises executive power through a Council of Ministers representing eleven Ministries. Tourism falls under the remit of the Ministry of Commerce, Industry and Tourism although, as a symptom of conflicts and weaknesses in the planning process related to tourism development, land-use planning policy is the responsibility of the Ministry of the Interior whilst environmental policy lies with the Ministry of Agriculture, Natural Resources and Environment (Loizidou 2004). Legislative power is exercised by the House of

Representatives, comprising eighty members (56 Greek Cypriot, 24 Turkish Cypriot, the latter remaining vacant whilst the 'Cyprus Problem' remains unresolved), again elected to five-year terms. However, despite this centralised system, much of the day-to-day administration of Cyprus is undertaken at the local level. The island is divided into six administrative districts, each headed by a District Officer, whilst within each district local administration operates through a three-tier structure of independent Municipal Councils, Improvement Boards and Village Commissions. Members of each of these are also elected to five-year terms.

Thus, for a nation with a population (approximately 670,000) equivalent to that of a medium-sized European city, Cyprus has a complex, multi-layered, democratic system of government, with a significant degree of authority delegated to the local level. There are, however, certain inherent weaknesses in the system. Firstly, formal structures for the implementation of policy at the national level do not exist. Instead, there is a reliance on informal contact and agreement between political and industry leaders as opposed to a formal consultation machinery; this, arguably, allows for political deals or favours that contradict or circumvent official policy whilst also permitting conflicts of interest.

Secondly, the multi-layered governmental structure, requiring elections at some level virtually every year, encourages policies and decision-making based upon short-term political motives. Thirdly, the multi-layered structure also places significant power in the hands of local municipal administrators, particularly mayors. With respect to tourism development, this gives the responsibility for land-use planning decisions, infrastructural development and other tourism-related activities to local politicians who, for electoral or other reasons, may not always make decisions in the wider regional or national interest.

Thus, as will be suggested below, tourism planning, particularly with respect to land-use policy, has suffered from a lack of formal implementation processes and the influence of local interests in decision-making. This in turn has allowed for excessive and, often, inappropriate developments, particularly in the accommodation sector, along the coastal strip as well as insufficient infrastructural developments. More generally, however, the outcome of the complex system of government in Cyprus is that a land-use policy specifically for the intensively used coastal strip (that is, a coastal zone management plan) is non-existent. As Loizidou (2004) observes, the number of involved authorities, the regular re-definition of coastal areas as tourism, as opposed to protected or agricultural, zones and, perhaps most significantly, the enormous financial interests involved in developing coastal areas for tourism, has resulted in a fragmented and erratic approach to coastal development.

This problem has also been undoubtedly exacerbated by the traditional importance of land ownership within Cypriot society. To Cypriots, land is valued both as a status symbol and for its potential economic return, whilst many Cypriots wish to pass on their land to their children as no inheritance tax is payable on it. Thus, the sale of land for hotel development, facilitated by the weak planning policy, has provided opportunities for rapid financial gain or to buy more land as an investment. Moreover, the Church has viewed tourism as an important source of income and, again taking advantage of lax planning laws, has either sold off or developed many prime sites for tourism development, particularly along the western coastline.

Interestingly, a coastal management project was initiated in Cyprus between 1993–1995, the aims of which were to establish a co-operative framework for managing the coastal strip and the establishment of general policy guidelines. However, although the outcomes of the project were accepted by the government, they were not introduced into planning procedures of the relevant authorities and, as Loizidou (2004) notes, the whole effort collapsed once the project ended. As a result, coastal tourism development has continued apace, contributing to the development of tourism which, as is now discussed, has increasingly focused on mass-market, summer sun tourism.

Cyprus: The Development of Tourism

As noted in the introduction to this chapter, forty years ago Cyprus was relatively unknown as a tourist destination – its tourism industry, comprising mainly small-scale, family run businesses located in hill resorts, was virtually non-existent (Ioannides 1992). By the 1990s, however, the island had evolved into a major Mediterranean destination, attracting almost 2.7 million tourists a year by 2001. Therefore, given its contribution to the island's economic growth, tourism in Cyprus may be considered a success story, particularly as there have been two distinct periods of development. However, as suggested above, the growth and scale of tourism development, the subsequent pressures on natural and human resources (Apostolides 1996) and the economic dependence on the tourism sector have long been considered unsustainable. As will be discussed, this has also been long recognized by the tourism authorities on the island and tourism policy has implicitly sought to achieve (albeit unsuccessfully) sustainable tourism development.

Interestingly, the challenges facing the tourism sector in Cyprus today have their roots in the first period of development up to 1974. During this period, the island experienced a rapid annual growth in tourist arrivals and receipts and, by 1973, tourism was already making a significant contribution to GDP (see Table 11.1). More importantly, perhaps, tourist activity underwent a fundamental spatial redistribution from the hinterland to the coast, with Kyrenia and Famagusta becoming the main tourist centres.

By 1973, these two resorts accounted for 58 per cent of the island's accommodation and 73 per cent of arrivals, with the Famagusta suburb of Varosha experiencing the greatest growth (Lockhart 1993). There, virtually the entire beach front had been developed for hotel accommodation. The composition of tourist arrivals also demonstrated an emerging dependence on the British market in particular, accounting for almost 44 per cent of arrivals by 1973, with the former West Germany and Scandinavia also important source markets.

The Turkish invasion in 1974 devastated the tourism industry. The great majority of existing and planned accommodation, as well as the international airport at

Table 11.1   Tourism growth rates in Cyprus, 1960–1973

	Arrivals / earnings				Rates of growth (%)		
	1960	1966	1971	1973	1960– 1966	1966– 1973	1960– 1973
Tourist arrivals ('000s)	25.4	54.1	178.6	264.1	13	25	20
Foreign exchange earnings (CY£m)	1.8	3.6	13.6	23.8	12	31	22
Contribution of earnings to GDP (%)	2.0	2.5	5.2	7.2			

Source: PIO (1997: 251); Ayers (2000)

Nicosia and many other tourist facilities, were within the occupied north and, hence, 'lost'. Nevertheless, as is well documented in the literature, tourism once again grew remarkably rapidly. The period between 1976 and 1989 witnessed an overall growth of 700 per cent in annual arrivals (Witt 1991) whilst, during the 1980s, tourist receipts grew by an annual average of 23 per cent compared with 8.3 per cent in Europe as a whole (CTO 1990). Such growth was driven partly by a no less rapid growth in accommodation facilities (Sharpley 2000), primarily in the coastal resorts of Paphos, Limassol, Agia Napa and Paralimni, the latter two resorts accounting for 39 per cent of the island's accommodation stock by 1990. By 2001, the island's bedspaces numbered 91,422, a sevenfold increase from 1980, with the great majority being located on the coast. Indeed, over the same period, the number of bedspaces in inland areas increased by just 11 per cent, from 3902 in 1980 to 4358 in 2001. The key indicators of tourism development in Cyprus since 1975 are provided in Table 11.2.

During the 1990s, arrivals were erratic and demonstrated relatively little growth – between 1994 and 1998, total growth of just 7 per cent was achieved, less than half the global rate. A significant increase was experienced between 1999 and 2001, primarily as a result of the short-lived popularity of Agia Napa as the clubbing centre of the Mediterranean. However, since 2001 arrivals have decreased alarmingly, as have receipts and, consequently, tourism's contribution to GDP. At the time of writing, the most recent figures suggest a slight recovery in arrivals in the first half of 2004 (a 4.7 per cent increase on the same period in 2003) although, most worryingly, tourist receipts continue to decline, indicating a decreasing average tourist expenditure.

Table 11.2   Tourism in Cyprus, 1975–2003 – key indicators

Year	Arrivals (000s)	Receipts (CY£mn)	Average tourist spending (C£)	Tourism receipts as % of GDP	Total licensed bedspaces
1975	47	5	n.a.	2.1	5685
1980	349	72	200	9.4	12,830
1985	770	232	299	15.7	30,375
1986	828	256	308	16.0	33,301
1987	949	320	334	18.0	45,855
1988	1,112	386	344	19.4	48,518
1989	1,379	490	350	21.7	54,857
1990	1,561	573	364	23.4	59,574
1991	1,385	476	343	18.4	63,564
1992	1,991	694	351	23.8	69,759
1993	1,841	696	379	21.4	73,657
1994	2,069	810	389	22.3	76,117
1995	2,100	810	383	20.5	78,427
1996	1,950	780	382	19.0	78,427
1997	2,088	843	393	20.4	84,368
1998	2,222	878	380	20.2	86,151
1999	2,434	1,025	400	22.0	87,893
2000	2,686	1,194	417	21.7	88,423
2001	2,697	1,272	441	21.6	91,422
2002	2,418	1,132	n.a.	18.4	94,466
2003	2,303	1,015	441	15.5	95,185

Source: CTO Reports; Republic of Cyprus Statistical Service; Ayers (2000)

A full discussion of tourism development in Cyprus is beyond the scope of this chapter. Nevertheless, it is important to identify the principal challenges facing the tourism sector on the island in the context of both tourism planning and EU accession. Beyond the erratic and, more recently, declining arrivals and receipts figures, the tourism authorities in Cyprus need to address the following issues:

- Tourism in Cyprus remains dominated by the British market. Despite attempts to diversify into new markets, in 2003 British tourists accounted for over 58 per cent of total arrivals, indicating an alarming dependence on a single market.

- As a result of restrictive air travel policies, two-thirds of all tourists travel to Cyprus on inclusive-tour or 'package' holidays. The number of independent tourists is on the increase (and will continue to do so as a result of EU accession), yet a significant proportion of tourism is controlled by tour-operators.

- Tourism to Cyprus remains stubbornly seasonal, with the peak summer season (July–September) accounting for 40 per cent of total annual arrivals.

- As observed above, accommodation development has been spatially concentrated in the coastal areas, whilst the latter half of the 1980s was notable for a rapid expansion in the provision of self-catering / apartment accommodation. Thus, although more recent growth has focused on the higher-graded hotel sector, the island's accommodation supply remains predominantly mid-range hotels and self-catering accommodation located in coastal resorts, underpinning the position of Cyprus as a summer-sun, mass-market destination. As a result, 93 per cent of all tourists stay in coastal resorts, which cover 37 per cent of the coastline of Cyprus. At the same time, the over-supply of accommodation has played into the hands of tour operators who are able to command heavy discounts, thereby reinforcing the mass market appeal of the island.

Figure 11.1a Famagusta, Northern Cyprus, a 1970s tourist resort. The rapid development of high rises was unsustainable since the beach does not get any sun even in the afternoon

Source: Richard Sharpley

Figure 11.1b Unlimited boom of tourist homes
Source: www.angelico.co.uk (28 January 2005)

Figure 11.1c Controversial hotel development right next to Akamas preserve
Source: www.conservation.org.cy (28.01.2005)

Importantly, underpinning the majority of these challenges has been the uncontrolled development of the coastline. Not only has the rapid expansion of accommodation facilities along the coast resulted in the loss of flora and fauna and agricultural land, increased coastal erosion, the introduction of 'architectural pollution', increases in air, water, ground and noise pollution and, generally, conflicts in land use, but also, as suggested above, the over-supply of accommodation has increased the dominance of overseas tour operators who, consequently, have been able to exert downward pressure on prices. As a result, tourism in Cyprus has become increasingly unsustainable: it is characterized by excessive coastal development, spatial concentration that minimizes the benefits of tourism in the hinterland, and mass market tourism that is becoming increasingly less profitable for the local tourism industry. What, then, have been the policy and planning responses on the part of the tourism authorities?

Tourism: Policy and Planning Responses

Given the importance of tourism to the Cypriot economy and the potential danger of unplanned, unconstrained tourism development, it would be logical to assume that official policies exist to guide the effective and sustainable development of tourism according to broader socio-economic and environmental development objectives. This is indeed the case. Since 1975, not only has tourism been identified as an important means of achieving economic growth and diversification in all national development plans, but also explicit proposals with respect to the scope, scale and nature of tourism development have been included in those plans.

Initial tourism policies focused on rebuilding the industry in the immediate post-invasion period, underpinned by various forms of financial support and incentives to encourage tourism development (Ioannides 1992). However, by the early 1980s it was evident that the re-development of tourism was too successful – the rapid development of accommodation facilities and the equally rapid growth in arrivals was not being matched by associated infrastructural development or the provision of ancillary tourist facilities. Moreover, planning controls to protect the environment were proving to be inadequate. Thus, in a sense, the successful re-development of tourism was threatening its own future.

Accordingly, a number of measures were introduced which, in effect, sought to limit the development of mass tourism on the coast, with 'the highest attention being paid to the protection and enhancement of the environment' (Andronikou 1986). These included a variety of financial incentives to encourage hotel and other tourism related development in the hinterland and the controlled development of luxury hotels in selected coastal areas. At the same time, marketing policy re-focused on attracting higher spending, 'quality' tourists, the purpose being to increase the value, rather than the scale, of tourism in order to reduce the island's increasing dependence upon the mass, summer-sun market. Efforts were also made to attract niche markets, such as conference/incentive tourism, special interest tourism and winter tourism, in order to address the problem of seasonality. In other words, the policy called for a more balanced, sustainable approach to tourism (Andronikou 1986).

In practice, the opposite occurred. As already noted, the 1980s as a whole witnessed a dramatic increase in annual arrivals, exacerbated from 1986 onwards by the introduction of charter flights. More specifically, between 1985 and 1990 the supply of accommodation on the island almost doubled, with 30 per cent of the increase attributable to new apartment accommodation in the coastal resorts. Nevertheless, in recognition of the fact that 'if the present course of [tourism] development is continued, it will in the long run have serious adverse effects on the competitiveness of our tourist product in the international market' (CPC 1989: 156), many of the policy objectives were embodied in subsequent economic development plans and tourism policies. These collectively reflected the Cyprus Tourism Organisation's (CTO) long-held objectives of:

- reducing the rate of growth in tourism development;
- upgrading and diversifying the tourism product, utilising the island's environmental and cultural attractions;
- spreading the benefits of tourism around the island;
- attracting more diverse, quality markets;
- increasing off-season tourism; and
- increasing the level of spending per tourist.

Attempts were also made to limit the environmental impacts of excessive development. A moratorium on new hotel building was imposed in 1989 (though this proved ineffectual given the large number applications approved prior to its imposition). Furthermore, in 1990, a Town and Country Planning Law was enforced, requiring all municipalities to submit local development plans for approval and, in particular, major hotel developments costing over CY£1 million to undergo an Environmental Impact Assessment.

Over the last decade, some – though limited – success has been realised in achieving these policies. For example, golf tourism has received a significant boost through the opening of two golf courses in the Paphos district, with three more under construction (though some would argue that, owing to perennial water shortages, golf tourism is an inappropriate development). The construction of six marinas has recently been approved, laying the foundation for developing potentially lucrative yacht-based tourism, and significant attempts have been made to develop agrotourism as a means of spreading the benefits of tourism away from the coast. Some sixty traditional rural properties have been redeveloped into tourism accommodation, whilst a number of villages have benefited from regeneration programmes. However, only about 450 bedspaces have been created and occupancy levels remain low (Sharpley 2002). Nevertheless, it has laid the foundation for attracting significant levels of Objective 2 funding from the European Regional Development Fund, of which a total of €28 million has been allocated to Cyprus for the period 2004-06 (Michael 2004).

Nevertheless, the island continues to attract a predominantly mass, summer-sun and lower spending market whilst, during the 1990s, the supply of licensed accommodation increased by 45 per cent, primarily in the coastal resorts. Efforts to reduce the seasonality of tourism have met with limited success and, most recently, average tourist spending has decreased. In other words, there has been a consistent failure to fully implement the desired tourism policy on the island or, more precisely, the CTO has been unable to manage effectively the development of tourism. There are two principal reasons for this.

Firstly, the CTO itself is a relatively powerless body, enjoying little statutory authority. For example, although it is able to license and grade new accommodation facilities, it is not in the position to decide whether such facilities should be built in the first place. Conversely, various political groups, such as the trade unions and the Cyprus Hotels Association (CHA), are relatively powerful. Thus, attempts to limit the growth of mass tourism in recent years have been thwarted by fierce lobbying on the part of the CHA to maintain occupancy levels, whilst rapid increases in wages have had a significant impact on profitability, productivity, investment and service levels.

Secondly, the lack of a cohesive and integrated coastal management plan, for the reasons outlined previously, has meant that the spatially concentrated exploitation of the coast for tourism development has continued. Equally, more general tourism development plans for Cyprus as a whole, embracing land-use planning, product development and enhancement, infrastructural development and improvement and so on have been difficult, if not impossible, to implement. Nevertheless, the authorities continue to pursue a quality / sustainable tourism development policy that, according to Michael (2004), is 'in absolute conformance with the European Community approach towards the sustainability of European tourism.' The current strategy for 2000–2010, focusing on the development of quality tourism, sets out a number of objectives (CTO 2001). These include:

- maximising the income from tourism (and, hence, its contribution to the national economy) through balancing a growth in arrivals with increasing visitors' length of stay and spending. The strategy proposes a 'value-volume strategy' which, by 2010, seeks to increase receipts to CY£1.8 billion based on arrivals of around 3.5 million;
- reducing seasonality, with the peak season's share falling from 40 per cent to 33 per cent of total annual arrivals;
- increasing competitiveness by re-positioning Cyprus as a tourism destination; in particular, less emphasis to be placed on sun-sea-sand tourism, whilst attention is to be focused on developing products, such as agrotourism, that are based around the island's culture, natural environment and people;
- attracting 'quality' tourists (defined in the strategy as older, better off, more culturally / environmentally aware and demanding flexibility, higher levels of service, better value for money) through more effective targeting and segmentation.

- in general, marketing the island as 'a mosaic of nature and culture, a whole, magical world concentrated in a small, warm and hospitable island in the Mediterranean at the crossroads of three continents, between West and East, that offers a multidimensional qualitative tourist experience' (CTO 2000: 33).

As the halfway point of this strategy's lifetime approaches, it is evident that many of its objectives will not be achieved although, following EU accession, Objective 2 funding will support the development of specialist, cultural and activity tourism in the hinterland, thereby potentially spreading the benefits of tourism whilst reducing pressure on the coastal environment. However, as the final section of this chapter now considers, the extent to which EU accession will counter the challenges currently facing the tourism sector in Cyprus remains uncertain.

Tourism in Cyprus: Implications of EU Accession

There is no doubt that, in general, Cyprus will benefit in many respects from membership of the EU, as of course will the other accession states. Furthermore, EU accession has presented a number of opportunities for the tourism sector in particular as the country aligns with EU legislation / directives and is able to benefit from structural funding. At the same time, however, these opportunities will also create challenges both for the tourism authorities and those agencies with wider planning and environmental responsibilities.

Firstly, European air liberalisation regulations will remove restrictions on air-carriers operating to and from Cyprus, implying that not only will more flights be available (particularly from key source markets) but also that fares will fall. In principal, this should result in increased numbers of tourists, particularly independent tourists (who tend to be more 'valuable' to the destination), and a subsequent fall in the number of inclusive tour/package tourists. This, in turn, may increase occupancy levels in hotels although a significant increase flights may be necessary, as will investment in the two main airports at Larnaca and Paphos. However, the assumption that low-cost air carriers will operate to Cyprus must be treated with some caution; for the major low-cost airlines in northern Europe, for example, Cyprus lies beyond their efficient operating range and, thus, the expected increase in numbers of tourists may not materialise.

Secondly, both second-home ownership by European nationals and time-share (the latter previously not permitted in Cyprus) will increase. Related to the increased accessibility to the island by air, this will again lead to an increase in tourist numbers, particularly as the renting out of private (second) homes will now be legal. Both second-home owners and time-share owners tend to benefit the local economy more than package tourists and it is likely that much development will occur away from the coastal areas. However, this will increase environmental and resource pressures in the hinterland, requiring more stringent planning controls, whilst a significant increase in European residents, as experienced in other destinations, may have social and economic implications.

Thirdly, EU structural funding will undoubtedly benefit the tourism-based revitalisation of rural areas, enhancing the existing agro-tourism programme, whilst environmental regulations in general, and schemes such as the Blue Flag campaign in particular, will continue to enhance the coastal environment. (In 2003, 39 beaches in Cyprus were awarded blue flags). Other EU directives regarding the safety of hotels, provisions for tourists and so on will also enhance the overall tourism product.

Fourthly, Cyprus currently suffers from a lack of employees in hotels combined with a relatively expensive workforce – the powerful trades unions in Cyprus have ensured that Cypriot hotel employees are relatively well paid. The free movement of labour within the EU will mean that new (cheaper) labour markets will be opened, although standards of service may fall, challenging the island's traditional reputation for high levels of hospitality. Unemployment amongst Cypriots may also increase as a result, although the more highly educated Cypriots are now free to work elsewhere in Europe, potentially fuelling a structural imbalance in the workforce.

Finally, the planned introduction of the Euro in a tourist destination that, by European standards, is already relatively expensive is likely to lead to further increases in costs. This, in turn, may decrease the demand for tourism in Cyprus, putting further pressure on the economic sustainability of the tourism sector.

Conclusion

There is no doubt that tourism has driven the development of Cyprus over the last forty years. However, it is also evident that the island's socio-economic development has not been achieved without cost, particularly with respect to the over-development of the coastal areas and subsequent pressure on natural and human resources. Moreover, as Cyprus has increasingly lost its competitive advantage, the contribution of tourism to the economy has diminished in recent years, potentially threatening the longer-term sustainable development of the island as a whole.

It is important to ask, therefore, whether the goal of sustainable tourism development, particularly as envisaged in the CTO's current strategy for developing quality, 'sun-sea-sand-plus' tourism, is a realistic or achievable objective. The evidence at present would suggest not. That is, although the most recent statistics point to an increase in visitor arrivals, the receipts from tourism continue to decline. Moreover, as the supply of accommodation increases further relative to demand, there will be greater downward pressure on prices, further reducing average tourist receipts and, consequently, the longer-term sustainability of the sector. As a result of the weaknesses in the planning system outlined throughout this chapter this trend is unlikely to be reversed, whilst local stakeholders have been unable or unwilling to limit the exploitation of the coast. Thus, it is likely that further developments will only be limited by the market, when the costs of investment significantly outweigh potential financial returns.

Equally, the position of Cyprus as a destination within the Mediterranean region cannot be overlooked. That is, although the current tourism strategy appears logical, it should not be viewed in isolation from the island's competitive position. Not only is Cyprus a relatively expensive destination within a price sensitive market (on average, a holiday to Cyprus costs 15 per cent more than elsewhere) but other destinations have, arguably, more diversity to offer. Both Turkey and Greece, for example, have a much greater supply of cultural assets, (yet, ironically, remain principally sun-sea destinations) whilst many other European destinations already offer the special interest holidays (golf tourism, cycling/walking tourism, marina tourism) that Cyprus is now striving to develop. Thus, as others have asked (Ioannides and Holcomb 2001), why should 'quality' or 'special interest' tourists choose to visit Cyprus?

A solution to the tourism problem lies, perhaps, in consolidating the island's current position. That is, as an established mass-market, summer-sun destination, attention should be focused on improving the quality of the resort-tourism experience in Cyprus through public investment in infrastructure and the coastal environment, thereby strengthening its position in the future. At the same time, economic development policies need to focus on reducing the island's dependency on tourism through promoting alternative economic sectors, thereby also potentially reducing the risk of out-migration of the island's highly educated workforce.

It is also highly likely that EU accession will result in greater economic stability in the future. However, it will not solve the current challenges facing the tourism sector, challenges which have, to a great extent, been self-imposed through inappropriate policies and, more significantly, the inability or unwillingness on the part of the government or tourism authorities to impose planning restrictions on tourism development. In conclusion, therefore, it is evident that, to address the current problems facing tourism and, indeed, to take advantage of the opportunities presented by EU accession, there is a need for proactive intervention on the part of the Cypriot government. Firstly, it needs to adopt a more appropriate policy for tourism development, replacing the rhetoric of sustainable tourism with a pragmatic acceptance of the island's principal markets. Secondly, there exists then undoubted need to intervene in the planning and development of tourism, particularly limiting any further development of the coastal areas. Otherwise, the tourism industry in Cyprus, and the island as a whole, is likely to continue to face an uncertain future.

References

Andronikou, A. (1979), 'Tourism in Cyprus.' In: E. de Kadt (ed.), *Tourism: Passport to Development?* New York: OUP, pp. 237–264.

Andronikou, A. (1986), 'Cyprus – management of the tourism sector.' *Tourism Management* 7(2), pp. 127–129.

Andronikou, A. (1987), *Development of Tourism in Cyprus: Harmonisation of Tourism with the Environment.* Nicosia: Cosmos.

Apostilides, P. (1995), 'Tourism development policy and environmental protection in Cyprus.' In: *Sustainable Tourism Development*, Environmental Encounters No. 32, Strasbourg: Council of Europe, pp. 31–40.

Ayers, R. (2000), 'Tourism as a passport to development in small states: the case of Cyprus.' *International Journal of Social Economics* 27(2), pp. 114–133.

Blake, A., Sinclair, M.T. and Sugiyarto, G. (2003), 'Tourism and EU accession in Malta and Cyprus.' <www.nottingham.ac.uk/ttri/pdf/2003_7.pdf>

Brey, H. (1995), 'A booming economy.' In: H. Brey and C. Muller (eds) *Cyprus*. London: APA Publications (HK) Ltd, pp. 92–93.

Cope, R. (2000), 'Republic of Cyprus.' *Travel & Tourism Intelligence, Country Reports*, No.4, pp. 3–21.

Coursaris, T. (1998), General Manager, Capo Bay Hotel, Protaras – personal communication.

CPC (1989), *Five Year Development Plan, 1989–1993*. Central Planning Commission, Nicosia: Planning Bureau.

CTO (1990), *Annual Report*. Nicosia: Cyprus Tourism Organisation.

CTO (2000), *Tourism Strategy 2000–2010*. Nicosia: Cyprus Tourism Organisation.

EIU (1992), *Cyprus*. International Tourism Reports 2, pp. 43–64.

Gillmor, D. (1989), 'Recent tourism developments in Cyprus.' *Geography* 74(2), pp. 262–265.

Ioannides, D. (1992), 'Tourism development agents: the Cypriot resort cycle.' *Annals of Tourism Research* 19(4), pp. 711–731.

Ioannides, D. and Holcomb, B. (2001). 'Raising the stakes: implications of upmarket tourism policies in Cyprus and Malta.' In: D. Ioannides, Y. Apostolopoulos and S. Sonmez (eds), *Mediterranean Islands and Sustainable Tourism Development: Practices, Management and Policies*. London: Continuum, pp. 234–258.

Kammas, M. (1993). 'The positive and negative influences of tourism development in Cyprus.' *Cyprus Review* 5(1), pp. 70–89.

Lockhart, D. (1993), 'Tourism and politics: the example of Cyprus.' In: D. Lockhart, D. Drakakis-Smith and J. Schembri (eds), *The Development Process in Small Island States*. London: Routledge, pp. 228–246.

Loizidou, X. (2004), 'Land use and coastal management in the Eastern Mediterranean: the Cyprus example.' www.iasonnet.gr/abstacts/loizidou.html.

Michael, P. (2004), *How is the Cyprus Tourism Organisation responding to Cyprus EU accession*. Paper presented to the 'Implications for the Cyprus Tourism Industry of EU Accession' Conference, Nicosia.

PIO (1997), *The Almanac of Cyprus 1997*. Nicosia: Press and Information Office.

PIO (2001), *About Cyprus*. Nicosia: Press and Information of Office.

Republic of Cyprus (2004), Statistics on tourism. Statistical Service, <www.mof.gov.cy/mof/cystat/statistics.nsf>

Sharpley, R. (2000), 'The influence of the accommodation sector on tourism development: lessons from Cyprus.' *International Journal of Hospitality Management* 19(3), pp. 275–293.

Sharpley, R. (2001a), 'Tourism in Cyprus: challenges and opportunities.' *Tourism Geographies* 3(1), pp. 64–85.

Sharpley, R. (2001b), 'Sustainability and the political economy of tourism in Cyprus.' *Tourism* 49(3), pp. 241–254.

Sharpley, R. (2002), 'Rural tourism and the challenge of diversification: the case of Cyprus.' *Tourism Management* 23(3), pp. 233–244.

Sharpley, R. (2003), 'Tourism, modernisation and development on the island of Cyprus: challenges and policy responses.' *Journal of Sustainable Tourism* 11(2+3), pp. 246–265.

Witt, S. (1991), 'Tourism in Cyprus: balancing the benefits and costs.' *Tourism Management* 12(1), pp. 37–46.

WTTC (2001b), Year 2001 TSA research summary and highlights: Cyprus. <www.wttc.org/ecres/a-cy.asp>

PART III
SPECIAL ISSUES IN URBAN PLANNING IN THE NEW EU MEMBER STATES

Chapter 12

The Increase of Urban Inequalities in Tallinn – Does EU Accession Change Anything?

Sampo Ruoppila

Introduction

In Estonia, as in many Central and Eastern European (CEE) countries, the housing reforms conducted after the collapse of state socialism have been characterised by a wholesale withdrawal of the state from housing provision (ownership, maintenance and allocation of housing) and an increase in homeownership. In the housing market that has emerged, the possibilities for households to access better housing, keep their present unit or improve their housing quality depend on their incomes. Since income inequalities have also become substantial (Mikhalov 2000), published research on post-socialist cities (e.g. Sailer-Fliege 1999; Sýkora 1999; Pichler-Milanovich 2001) unanimously agrees that one aspect of the transition is increasing residential differentiation, i.e. more unequal distribution of social groups within a city (as well as within an urban region).

This chapter analyses the development of the housing sector in Tallinn, the capital of Estonia, from the perspective of residential differentiation. The level of residential differentiation was low at the end of the socialist era, but since the beginning of the transition it has gradually increased. Moreover, as the chapter will show, there are no policy constraints for that increase, and the economic and housing problems of the low-income groups are altogether disregarded in housing and social policy. As this stands in contrast to policies applied in most Western European countries, which actively aim to hamper social inequalities and segregation in their housing markets, the question arises of whether EU access can be a trigger for a change in policies in Estonia.

The chapter is divided into four sections. The first introduces the context of the change: the introduction of liberal economic, social and housing policies as a part of the transition in Estonia. In the second section, the development of the housing market and changes in the pattern of residential differentiation are analysed. The third section focuses on contemporary Estonian housing policy, its impact and limitations. In the fourth and final section, a commentary is added on how the EU may affect policies related to the housing sector in Tallinn – and other urban areas of the Eastern part of the EU.

Context: Transformation Processes in Estonia

The Societal Transformation

Since regaining its independence in 1991, Estonia has carried out radical reforms that have transformed it in a mere decade from a strict planned economy to one of the most liberal economies in the world.[1] In large-scale privatisation, capital assets were shifted to private owners, and subsequently private control over the economy was increased. A guideline of economic policy has been the creation of optimal conditions for entrepreneurship and economic growth with the principle that state intervention in the economy should be limited. The macro-economic policy is based on the fixed exchange rate of the Estonian Kroon (EEK) to the D-Mark, which was set in 1992, a balanced budget and a low level of state debt. Price regulation was quickly abolished (with the exception of energy, water, health services and postal services) as were trade barriers and limitations on the movement of capital (Eamets 1999). Taxation is very enterprise friendly: companies must pay corporate income tax (26 per cent) only on distributed profits. A flat personal income tax (also 26 per cent) favours individuals with high salaries. However, the raising of the level of tax free income to EEK 16,800 (€ 1073) per year from the beginning of 2004 helps low-income groups.

Formulations of social policy have also been (neo)liberal, with an emphasis laid on the freedom and responsibility of individuals. Basic education is still provided without any fees and statutory health insurance covers all residents, but the social safety net has been lowered in terms of social benefits. The monetary level of the social safety net is so low that it hardly guarantees survival: an unemployment benefit (EEK 400 / €26 a month, paid only for six months) and the national pension (EEK 900 / €63) provide an income lower than the subsistence minimum (EEK 1,500 / €96, calculated by Statistics Estonia in spring 2004). The subsistence benefit, which is supposed to help low-income households to cope with living costs, guarantees a net income (EEK 500 / €32 for the first applicant and EEK 400 / €26 per additional member of a household after the payment of fixed housing costs) that is even less than the cost of the minimum food basket alone (EEK 700 / €45 per person)! Moreover, the monetary level of the subsistence

---

1   For instance, in the 2004 index of economic freedom (Miles et al. 2004), issued by the Heritage Foundation and the Wall Street Journal, Estonia ranked sixth after Hong Kong, Singapore, New Zealand, Luxembourg and Ireland, above countries such as the UK, Denmark, Switzerland and the US.

benefit has remained the same for the last seven years (since 1997) despite the fact that prices and the general income level have risen considerably.[2]

On average, the standard of living for Estonians has improved since 1994. Estonia's GDP per capita reached €5,942 in the year 2003. In the second quarter of 2004, the average monthly salary was EEK 8,999 (€575) in Tallinn and EEK 7,417 (€474) in the whole of Estonia, and the average old-age pension was EEK 2,129 (€136). However, income disparities are high. In Tallinn, the difference between the disposable income of the richest 10 per cent and the poorest 10 per cent of households was thirteenfold in 1999 (Ruoppila and Kährik 2003: 56). The proportion of low-income people is substantial: roughly one fourth of all Estonian households live in poverty (Social Trends 2 2001: 70). Families with many children, single parent families, and households whose family head (i.e. the biggest earner) is either unemployed, has a low level of education or is a pensioner are the ones which most commonly have economic difficulties.

The Transformation of the Housing Sector

Concurrently with changes in the economic and social sectors, the Estonian government launched a privatisation programme and a number of other reforms in order to replace the former public allocation system of housing with a market mechanism.[3] The transfer of the former public rental housing to private ownership was particularly extensive; in Tallinn, the share of privately-owned dwellings rose from 25 to 94 per cent between 1993 and 2000 (Tallinn arvudes 1992; Living Conditions 2000). The transfer consisted of the restitution – i.e., the return of dwellings nationalised by the Soviet authorities to their former owners or their legal heirs – and the privatisation of dwellings in state or municipal ownership. Of the dwellings constructed before 1945, in Tallinn altogether 41 per cent were returned (Omanikele tagastatud... 1998: 27; Tallinn arvudes 1992: 48). Most of the dwellings belonging to the municipality were privatised and transferred to the sitting tenants. The people who lived in the restituted houses could not buy their flats, but became tenants of the private landlord to whom the ownership was returned. The public rental dwellings that were not privatised (six per cent of the stock) are mostly of a low standard and primarily occupied by socially disadvantaged groups (For more details, see Estonian Human... 1997: Ch. 4; Kährik 2000).

In the privatisation of housing, transactions were conducted with vouchers, which were given to all individuals who had permanently lived and worked in Estonia for at

---

2   The monetary level of the social benefits mentioned above was checked as of summer 2004.

3   During the Soviet era the state had a central role in the production, ownership and distribution of housing. The dominant form of tenure was public rental housing (76 per cent in 1992), which was distributed via the trade unions. The provision of other tenures, owner occupation (10 per cent) in the form of single-family housing and housing co-operatives (14 per cent), were strictly regulated by the state, and were complementary to the provision of public rental dwellings.

least five years. Thus, a household's financial resources or ethnic status did not affect its ability to buy the dwelling. Moreover, differently from many Central European states, the flats were transferred to private ownership without the incurrence of debt. There were two overriding motives for tenants to buy their flats. On the one hand, the value gap between the cheap purchasing price and the higher market price (of any kind of flat) was attractive. On the other hand, the tenants feared that unless they bought, they would eventually lose the right of occupancy. Many followed the herd and became homeowners without understanding the consequences. A popular misunderstanding was, for example, that it would be immediately cheaper to live in an owner-occupied flat since [the formerly heavily subsidised, all housing costs incorporated] rent would no longer be payable (Kõre et al. 1996: 2155).

Responsibility for the housing management of the privatised multi-family units was transferred to Homeowners' Housing Management Associations established in each unit. The misconceptions of cheaper living were quickly shattered as energy prices increased to the market level and the subsidies that had defrayed the maintenance charges were also cut. Moreover, any extensive renovation needed after previously deferred maintenance (i.e. repairs neglected during the socialist era) is to be financed by the homeowners within the next decades.

Housing allowances were initially introduced to offset increased housing costs for both owner-occupiers and the tenants. However, in order to reduce governmental spending on housing, they were combined with a 'supplementary benefit' into one unitary measure 'subsistence benefit' in 1997 (in Tallinn in 2001). It is provided only to households with incomes far below the subsistence level.

The city of Tallinn continued to provide public rental dwellings, although no new public rental housing was constructed in Tallinn from 1993 to 2000. (The only thing the city did was to renovate single blocks as 'social housing' for the elderly and the disabled.) Most of the old stock is (unspecified) 'municipal rental housing'; 'social housing' exists as a sub-category. Eligibility for municipal rental housing is based on three criteria. The first is homelessness due to an accidental event such as fire or expropriation of property, and also applies to released prisoners and orphans (if they are permanent residents of the city). The second applies to tenants in restituted houses regardless of their income, and the third to tenants in uninhabitable municipal rental houses and elderly, disabled or poor families who need social assistance. Only households with special needs are eligible for social housing: the elderly and disabled, released prisoners, orphans and low-income families in need of frequent social assistance including custody of their children. What is notable – and different from most European countries – is that poverty alone does not make a household eligible for municipal or social rental housing.

There is one group in particular that Estonians (and the press) associate with the housing problem: the tenants of restituted houses – or 'forced tenants' (sündüürnikud), as they are called in Estonian. Not only were they denied the chance to become homeowners, they are also regarded by the new owners of the properties as obstacles in terms of their right to use the property, and their tenancy has repeatedly been challenged on these grounds. Because the problem was caused by the ownership reform, the tenants have enjoyed the special protection of the government. Their rents have been regulated – and eviction for reasons other than unpaid rent has been possible only if alternative rental accommodation is found. They are entitled to municipal rental housing, but given the limited stock, only a few have been thus accommodated. In 1998, a total of 23,470 Tallinnites lived as tenants in restituted housing (Omanikele tagastatud... 1998).

During the 1990s the housing policy agenda was characterised by the withdrawal from the socialist system. In the state policy the emphasis was on establishing an institutional and legal framework that would enable individuals to operate in the housing market through loans. At the municipal level where privatisation was implemented that was the main concern. Comprehensive policy programmes to support the renewed housing system have been introduced only in the 2000s.

The Changes in Residential Differentiation

The Starting Point: Inherited Urban and Social Structure

The residential districts of Tallinn are shown in Figure 12.1. The centre of Tallinn consists of a medieval old town (number 1 on the map), a small modern central business district, and a mosaic of distinctive housing areas (2). What distinguishes the city from other European capitals is that even in the very core of the city, many areas contain plenty of housing built out of wood, and some old suburbs such as Kalamaja (4) and Pelgulinn (5) are almost completely wooden housing districts. This is a reflection of Tallinn's 'colonial' history as a peripheral industrial city before the First World War. Neither the Germans nor the Russians, who owned most of the manufacturing industries at that time, invested in Tallinn, but rather put their resources into St Petersburg or German cities (Bruns 1993a: 91–93). The building code was not updated either, which enabled Estonian landlord-developers to use traditional and cheap building materials for the tenements they provided for the growing urban population. Modernisation changed the face of some areas during the era of independence (1918–40), although wooden housing constructions continued in non-central locations. Central areas demolished in WW II were redeveloped at the beginning of the Soviet era (1944–91) to conform to the Stalinist classic style. Otherwise the historical urban structure survived relatively unscathed, although the buildings gradually deteriorated due to a lack of maintenance. Housing construction started to increase in the 1960s, but the emphasis was on constructing new areas on the outskirts of the city. The demand for the new units was high, given the extensive population growth, speeded up also by immigration from Russia and other Soviet republics. During the Soviet era, Tallinn tripled its population to 479,000. Housing development was concentrated in three housing estate districts, Mustamäe (12), Väike-Õismäe (13) and Lasnamäe (15), where altogether half of Tallinn's population live. The single-family housing areas (10, 11, 14, 16, 17, 18, 19, 21) that emerged in the suburbs during the inter-war period were also extended during the socialist era (especially in the 1950–1960s) and are expanding now.

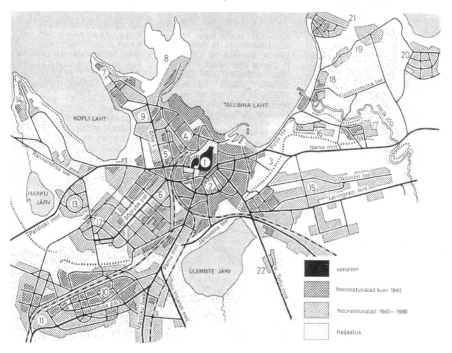

Figure 12.1   Districts of Tallinn
Source: Bruns 1993b: 15

The level of socio-occupational residential differentiation was low at the end of the socialist period. A survey conducted in 1981 (Raitviir 1990) showed that the proportion of white-collar employees (with higher or college education) and blue-collar workers in the working population varied little across the 23 districts used in the study. However, some differences appeared. Socio-occupational status was relatively the highest in two central districts that have a concentration of good quality blocks constructed out of stone built in the inter-war period and the 1940s and the 1950s, as well as in single-family housing districts in the northeast. (To build a single-family house required savings and social relations, which those in higher occupational positions were more likely to have.) The socio-occupational status was the lowest in two traditional workers' neighbourhoods: the northern part of the Kopli peninsula (Pelgurand, Kopli) and semi-industrial districts in the south side of the city (For a more detailed account, see Ruoppila and Kährik 2003: 53–55; Ruoppila 2004).[4]

---

4   Ethnic residential differentiation was more apparent given that non-Estonians were privileged over Estonians in the distribution of new public rental dwellings. Consequently, non-Estonians were over-represented in multi-family housing constructed in the socialist era, whereas pre-socialist housing, and especially single-family housing, was more often occupied by Estonians (Org 1989). Nonetheless, neither socio-occupational nor ethnic differentiation

The Emerging Housing Market

Due to the extensive privatisation, a characteristic of Tallinn's housing market is a very large share of owner-occupied housing. The private rental market is small, with only a few per cent of the housing stock added to the original tenancies in restituted houses.[5] In the 1990s the level of residential mobility was low: only five per cent of Tallinnites moved in 1998 and four per cent in 1999 (Ruoppila and Kährik 2003: 66). The mobility rates were higher for the highest and the lowest income quartiles than for the middle-income groups. For the wealthier Tallinnites new possibilities opened up in the housing market, which explains their higher mobility, whereas the higher mobility of the poor is rather to be explained by 'push factors' such as housing costs exceeding their financial resources and the consequent necessity to find more affordable housing. The very low mobility rate of the 2nd income quartile (2 per cent) can be explained by the ability of these households to cover the maintenance costs of their dwelling, and the lack of resources to buy a new dwelling (ibid.).

For most buyers their old dwelling, usually the privatised debtless flat, is an important equity. Most of those who relocate to a more expensive unit or buy their first dwelling take a housing loan to finance their purchase. Commercial housing loans have been available since 1996 and the system is well developed. However, only about one fourth of Estonians have a high enough income to be able to take the housing loan (as estimated by the commercial bank Hansapank in August 2004). Even a household with two average income members may find it difficult to enter the housing market if they are buying their first dwelling in the capital city. This is the situation of many young households as well as those wishing to move to Tallinn from regions with significantly cheaper real estate prices. Kährik (et al. 2003: 33) found that in the year 2000 as much as one quarter of people aged 30 were still sharing dwellings with their parents (or grandparents), presumably because of the difficulty of gaining access to housing.

The majority of dwellings sold in the housing market are units constructed during the socialist era or before. As shown in Table 12.1, dwellings constructed since 1991 form yet a small part of the housing stock. This is because the low purchasing power has limited the demand for the privately developed units and no new public rental housing was constructed between 1992 and 2000. The decrease in population has also favoured the fulfilling of housing needs through redistribution of the existing housing stock. However, in the census (2000) the number of households still slightly exceeded the number of dwellings.

---

was significant in the high-rise housing estate districts of Mustamäe and Väike-Õismäe. Only the Lasnamäe district has an ethnically Russian bias.

5   My broad estimation is based on stray survey results about the size of the rental sector in total, and the data about the share of the public rental sector and the number of restitution tenants.

Tallinn's population decreased 16 percentage points from 1989 (479,000) to 2000 (400,000) due to a negative natural increase in population and emigration of part of the Russian speaking population. Tallinn also lost population to its neighbouring municipalities due to suburbanisation, but gained the same amount of population from elsewhere in Estonia (Tammaru 2001). Since 2000, Tallinn's population figure has remained around 400,000.

Table 12.1   Tallinn's housing stock by year of construction (number of dwellings) as of March 2000

	Before 1919	**1919– 1945**	**1946– 1960**	**1961– 1970**	**1971– 1980**	**1981– 1990**	**1991– 3/2000**
Blocks of flats	5723	12479	11875	37619	42241	36627	4079
Single-family houses	382	2577	3231	1998	403	384	1022
Other small residential buildings	203	1589	675	408	85	117	312

Source: Census 2000

Differentiation Processes in the Housing Market

The average prices of flats and single-family housing across districts are given in Tables 12.2, 12.3 and 12.4. These prices indicate the current appreciation of different districts.

Further, as the access to housing is determined by a households' cash resources, prices may be taken as a rudimentary indicator of the current trends of residential differentiation; presumably more expensive housing areas draw on average those with higher incomes (or more wealth) and those with smaller incomes (or less wealth) have to settle in cheaper housing areas. (€1 = EEK 15.65.)

Likewise, as new and rehabilitated housing has been accessible only to the relatively wealthy, the spatial concentration of this housing can be taken as an indicator of the spatial concentration of better-off households. Following this indicator, the wealthy are (further) concentrated in and around the city centre, and in the garden suburbs and the adjoining countryside.

The new blocks of flats initiated by commercial developers are located mostly in and around the city centre. The majority of these new blocks of flats have replaced modest wooden housing and thus promote the expansion of a modern stone city. However, within two last years, i.e. after the map was drawn, construction of blocks of flats has also increased in garden suburbs, but still not in old housing estate districts. As to the new single-family housing areas, in November 2002 there were around 50 commercial developments finished or under construction in and around Tallinn, with a total volume of 2500 lots planned. The developments are concentrated on Tallinn's northeast and northwest shores, expanding the existing single-family housing areas. The biggest projects (with more than 200 lots) have so far been located on the edges of the city within the administrative borders. Nonetheless, the first wave of sprawl has emerged as the residential areas have expanded into the adjoining countryside (Ruoppila 2002).

Table 12.2   The sales prices of used standard two-room flats, EEK / m², June 2004
*Refers to the dwelling, not necessarily the rest of the house.

		Renovated*	Not renovated
Old Town		From 22,000	From 18,000
City centre		From 14,000	From 12 500
Kadriorg	Stone houses	15,700 – 20,000	10,500 – 14,000
	Wooden houses	13,500 – 14,000	10,500 – 12 500
Kalamaja, Pelgulinn		From 10 500	From 9,000
Kopli		From 10,000	From 8,000
Mustamäe		12,000 – 13,000	10,500 – 11,000
Haabersti (Väike-Õismäe)		12,000 – 13,000	10,500 – 11,000
Lasnamäe		9,700 – 11,000	9,000 – 9 400
Lilleküla, Kristiine		From 13 500	From 11,000
Nõmme, Pääsküla, Laagri		From 12,000	From 10,000

Source: Market review of the real estate company Arco Vara

Table 12.3   Prices of new flats, EEK / m², June 2004

City centre	15,000 – 25,000
Pirita garden suburb	15,000 – 25,000
Elsewhere in outer city	12,000 – 17,000

Source: Real estate market reviews, newspaper advertisements

Table 12.4   Sales prices of single-family housing, EEK / m², June 2004

*Excluding villas from the inter-war period as well as other architecturally outstanding houses

	Contemporary, new	Built in the 1980s and the 1990s	Built before the 1980s*
Nõmme, Hiiu, Pääsküla	13,000 – 18,000	10,000 – 14,000	9,000 – 14,000
Maarjamäe, Pirita, Pirita-Kose, Merivälja	13,000 – 18,000	8,000 – 12,000	8,000 – 11,000
Kakumäe	12,000 – 14,000	8,000 – 12,000	8,000 – 10,000
Lilleküla, Veskimetsa	10,000 – 14,000	6,000 – 10,000	6,000 – 10,000

Source: Market review of the real estate company Arco Vara

Developers have also been active in the rehabilitation of the pre-socialist housing stock. Given the restitution of property, the developers have been able to acquire ownership of a whole house, followed by the displacement of the tenants, full renovation and sales (or rentals) to new, wealthier inhabitants. These developments have been concentrated in neighbourhoods that were already highly appreciated before the socialist era, including the old town, city centre, Kadriorg and the garden city of Nõmme. In the 1990s it was primarily the buildings constructed of stone that were renovated, but in the 2000s the rehabilitation of wooden housing has also increased. After decades of neglect, some parts of the wooden housing districts, especially in Kalamaja, have even started to show signs of gentrification. The usual methods of displacing the tenants have included acquiring an alternative flat or offering them money for moving out and giving up the status of a 'tenant in a restituted house'.

As to the secondary housing market, other than fully renovated houses, the two economic extremes appear as well. On the one hand, only those who can afford the above average prices can buy dwellings in the city centre, Kadriorg and the most popular single-family housing areas. On the other hand, those who have the fewest choices, including poor owner-occupiers changing to cheaper flats as well as displaced tenants of restituted houses (to whom an agent used by the developer acquires an alternative flat) are most likely to end up in deteriorated parts of wooden housing districts (Kopli, Kalamaja, Pelgulinn) or in Lasnamäe. Comparing the economic background of those who had moved between 1991 and 1999 and the housing types that they had moved into (flats or single-family houses, with or without all the basic facilities), Ruoppila and Kährik (2003: 67) were able to show that in the 1990s mobility had increased the share of high-income households in single-family housing with all facilities (i.e. the single-family housing areas) and the share of low-income people in flats without all basic facilities (i.e. wooden housing districts, old workers' housing areas).

It is noteworthy, however, that most of Tallinn's housing areas fall into a broad and mildly differentiated mid-price category. For instance, this is the case for most of the housing in the three large socialist housing estate districts. Dwellings in them are still considered 'basic' units and no stigma is attached to them. This is also interesting as since the mid-1990s (e.g. Kinnisvara ekspert 1997) the relative price differences between the districts have rather decreased than increased, and in none of the areas have the prices dropped altogether. My interpretation is that because of the lack of affordable alternatives, people have tried to do the best they can with the existing housing stock, including the not so attractive parts of it. No dramatic residential differentiation can hardly be said to have taken place between the mid-priced housing districts in the 1990s.

The Pattern of Residential Differentiation After One Decade of Transition

Tallinn was not yet divided into large rich and poor areas at the end of the 1990s, i.e. after the first decade of transition. Surveys run in 1996 (Loogma 1997) and 1999 (Ruoppila and Kährik 2003) as well as the census data (2000) on the distribution of people with different levels of education (ESA 2002) confirmed the prevalence of a rather low level of residential differentiation between Tallinn's eight administrative city districts. Yet the socio-economic position of the population was higher than average in the single-family housing districts Pirita and Nõmme, followed by the districts of Central Tallinn (the city centre and its environs) and Haabersti (including the Väike-Õismäe housing estate and the Kakumäe single-family housing area) and was the lowest in the district of Northern Tallinn (including Kopli, Kalamaja, Pelgulinn, Pelgurand etc.); the socio-economic position was also somewhat lower in Lasnamäe.

Despite the low level of residential differentiation at the district level, Ruoppila and Kährik (2003) found evidence of social polarisation related to housing quality within the districts that were developing more rapidly due to construction and renovations. This was clearly the case in the district of Central Tallinn. In addition, in the old garden city of Nõmme the highest income group was concentrated in single-family houses with all facilities. The result suggests that pockets of wealth and poverty were developing within an otherwise mixed socio-spatial structure.

Estonian Housing Policy in the 2000s[6]

Estonian housing policy moved to a 'post-privatisation' phase only after the new millennium as both the state (MKM 2003) and the city of Tallinn (TLV 2002) published their new housing policy agendas to support the renewed housing system. The central government has the roles of coordinator, strategic planner, and legislator, but local governments are responsible for the housing issues. In practice, both the state and the municipalities have drawn and implemented their own housing policy programmes and measures, yet their focus and ideas are quite similar. On the basis of their purposes, the programmes can be divided into three categories. They focus on (1) increasing access to home ownership, (2) speeding up the renovation of owner-occupied multi-family housing and (3) developing municipal rental housing. All the state measures mentioned below are distributed through the Estonian Credit and Export Guarantee Fund (KredEx) and implemented by the local municipalities.

Measures Increasing Access to Home Ownership

Housing policy programmes that fall into this category are designed to support households individually solving their housing question by buying a dwelling. The state is attempting to increase the number of housing loans provided by commercial banks. It secures the loan, and thus enables borrowers to make a smaller down payment (decreased from 34 per cent to 10 per cent) and gives them a longer payback period. This measure is meant to benefit first-time buyers, and eligibility is restricted to families with a child or children under 16, young professionals (with academic or specialist education who are under 35) and tenants of restituted houses. According to data provided by the commercial bank Hansapank, 13 per cent of the housing loans issued in the Tallinn area in 2003 were guaranteed by the state.

The City of Tallinn has a programme which is much smaller in scale to build owner-occupied housing through a public-private partnership. The idea behind this programme is that the city may be able to restrain migration from Tallinn to its neighbouring municipalities by increasing the supply of new affordable small-scale housing within the city's administrative borders. However, although the programme was launched in 2001, by 2004 only one small single-family housing area (67 dwellings) had been completed. The role of the city in the partnership was to draw up a land-use plan and to provide building land and infrastructure. Because of the subsidy, the prices of the dwellings were roughly 20 per cent below the market price. Nevertheless, there were no restrictions on who could buy them.

Measures Supporting the Renovation of Blocks of Flats

Multi-family housing units handed over to sitting tenants have not been renovated on a mass scale until recently. A crucial factor here was the introduction of renovation

---

6   This section draws from Ruoppila (forthcoming).

loans given to Homeowners Housing Management Associations (henceforth HHMAs) by commercial banks. Nevertheless, the renovations have proceeded very slowly, which is why both the state and the city of Tallinn have introduced their own programmes to speed them up. Renovations entail improvements that enhance the technical quality of the buildings, such as new heating systems, better insulation, and new roofs.

The state programme provides two non-refundable grants. A 'general grant' supports all renovations that enhance the quality of the building, and covers 10 per cent of the costs. A 'technical grant' covers 50 per cent of the costs of a technical inspection of the house that is made prior to the renovation. The city of Tallinn subsidises the interest on the renovation loans given by the commercial banks, thus making the subsidised loan more affordable than a normal loan for the HHMAs. The grant given by the state and the loan-interest subsidy provided by the city have been popular, and demand has exceeded supply: between April 2003 and June 2004, 136 HHMAs in Tallinn were given state renovation grants (general grants), 79 were given subsidised loans in 2003 and 77 in 2004. The subsidised loans for 2004 were sold out by the beginning of May.

The support provided by the state and the city functions as an incentive for HHMAs to take renovation loans. Their ability to do so depends on the paying ability of the inhabitants, which favours those in which middle- and upper-strata households form the majority.

As both grants and interest subsidies are open to all HHMAs regardless of what kind of multi-family houses they manage or where these houses are located, one would expect more grants and subsidised loans to have been assigned to rapidly upgrading parts of the city where better-off households are settling. This assumption could be tested using the data obtained on the distribution of the 'general grants' in Tallinn in 2003 and 2004. However, contrary to the expectation, they had been assigned across the districts and to different kinds of multi-family housing areas. Nevertheless, as expected, the monetary level of the grants (i.e. how much the 10 per cent of the subsidised renovation was in EEK per sq. m.) was higher in the favoured and gradually gentrifying neighbourhoods. The subsidies were larger in the districts of Central Tallinn and Nõmme, as well as in the inner city neighbourhoods of Kalamaja and Pelgulinn (in Northern Tallinn), indicating that the HHMAs in these areas were conducting more extensive renovation projects (i.e. doing several things at the same time, such as re-plumbing, façade renovation, re-roofing and improving the heating systems). According to one civil servant working in the city administration, the pattern of the subsidised loans was similar; they were also distributed across Tallinn, but they were taken up most actively by HHMAs in the Central Tallinn and Northern Tallinn districts (the Kalamaja and Pelgulinn neighbourhoods in particular in the latter district), and there also the amount of loan per square meter was higher.

One consequence of this increasing amount of renovation is the displacement of poor households. Decisions to renovate and to take a loan for that purpose are made in HHMAs by a simple majority of votes (all homeowners have one vote). Thus in blocks in which the wealthier households are in the majority, the renovation can start and the poor have to sell their flats and move. Housing associations have a legal right to apply for the eviction of owner-occupiers who cannot pay their share of the housing costs, including the costs of renovation. The relocating poor have to move to a more affordable flat. This is a mechanism that strengthens the relationship between householders' financial circumstances and their housing quality, and leads to residential differentiation between renovated and unrenovated blocks and areas.

The pressure to begin renovating regardless of the social effects of displacement is currently the highest in the appreciating centrally located neighbourhoods. It seems to be a case of the increase in renovations leading to the relocation of people according to their ability to pay. Presumably, this will trigger a whole new level of residential differentiation.

The Development of Municipal Rental Housing

In 2001, after a break of one decade, the city of Tallinn started a development programme of new public rental housing. By February 2004, twelve new houses with altogether 668 new municipal rental dwellings had been built, and the plan is to continue at the rate of 400 dwellings a year until 2008 (TLV 2003). In terms of scope and cost, this is currently the most extensive housing-policy programme being implemented by the city of Tallinn. In principle, the state provides a subsidy of up to 50per cent of the costs of development of municipal rental housing, but in the case of Tallinn the proportion of state finance has been much smaller, about five per cent in 2004.

Contrary to the idea it evokes in the mind of a Western observer, the municipal rental housing programme in Tallinn is not about social housing for the low-income population, but it is about relocating the remaining tenants of restituted houses to municipal flats. It is an attempt to do away with the transitional tenure of 'tenant in a restituted house', which has been considered a major headache in Estonia since the beginning of the property reform. The fact that it is promoted purely as a social programme benefiting tenants is sheer rhetoric. It is obvious that property owners also benefit as tenants are displaced at the expense of the municipality and the state.

The new municipal rental dwellings are allocated solely to the tenants in restituted houses who are entitled to rental accommodation regardless of their income. Other groups eligible for public rental housing, mostly low-income people, only have access to the old residual housing stock and not to the new dwellings. Therefore, the programme does not improve the housing access of most low-income people.

Housing Problems of Low-Income People Continue to be Neglected

The problem with the housing policies in Estonia is not what has been done, but what has not been done. Most importantly, the issue of social justice is neglected:

the problems of households with less-than-average incomes are not being addressed by the state or by the city of Tallinn.

The majority of households (three out of four in the whole of Estonia, according to the estimates of the commercial banks) do not qualify for housing loans even if they get additional support (security) from the state. It is difficult for low-income households to pay back renovation loans taken by the HHMA, and the poor are already struggling to meet maintenance costs. Poverty or low income alone does not make a household eligible for municipal or social rental housing. New municipal rental accommodation only caters for tenants from restituted houses, and most low-income households are thus left to cope with the means they have. Furthermore, their vulnerability to homelessness is increased by inadequate housing allowances. The only support available is a subsistence benefit, which is paid only to households with incomes far below the official subsistence level, and which is only sufficient to cover the housing costs of tenants in municipal rental, social rental and restituted housing (the original tenancies), who pay regulated rents.

Homelessness, which emerged as a new phenomenon in the mid-1990s, is still surprisingly rare considering the disregard for the needs of the poor. The total number of homeless people, according to the expert evaluation of social workers, is 3000–3500 people (0.25–0.30 per cent of the population) in Estonia, of whom 2000 are located in Tallinn (0.5per cent of the population). Unsurprisingly, the amount is expected to increase (Kährik et al. 2003: 41).

While the current housing policies increase the choices of households who are able to buy or to improve their properties with the help of loans, the lack of measures designed for low-income groups hastens their relocation. If their housing costs exceed their financial means, they are obliged to move to accommodation of a smaller size or inferior quality, or to a cheaper area. In these circumstances it is inevitable that the relationship between household income (or wealth) and housing quality will strengthen, which in turn will increase socio-economic residential differentiation across urban neighbourhoods in Tallinn.

Does EU Accession Make a Difference?

This chapter has shown that Tallinn is currently at a crossroads. On the one hand, the low level of residential differentiation that was the starting point after the socialist period has not *yet* vanished. On the other hand, residential differentiation is increasing and the very liberal social and housing policies allow an unrestrained increase of inequalities. But does EU accession have an(y) impact on the Estonian housing policy or housing sector? Are there any grounds for optimism that the social side will be strengthened?

Because the housing policy is the exclusive responsibility of the member states, EU access does not have a *direct* impact on national housing policies. Nonetheless, EU policies do include actions that will have an *indirect* impact.

Figure 12.2a Gentrification in Kadriorg
Source: Sampo Ruoppila, 2001

Figure 12.2b J. Kõleri Street, wooden tenement housing
Source: Kertu Kurist, 2003

Figure 12.2c Renovated multi-family house with new ridge roof in the Mustamäe housing estate district

Source: Andres Kurg, 2004

In a recent analysis on the future of the social policy of the European Union and its implication for housing, Mark Kleinman (2002) distinguishes between three ways in which housing will continue to have a role in the EU at the supranational level. First, housing policy is involved in economic policy. European institutions are concerned with issues such as the regulation or liberalisation of mortgage markets, the level of spending on housing both by households and by governments, housing finance and its effects on public sector balances, and competition in the construction industry. Secondly, there is already a network of institutions and lobby groups around the activities of the EU in the housing sphere and closely related spheres such as social

Figure 12.2d Old workers' tenements in Kopli Lines
Source: Triin Ojari, 2004

inclusion. These include homelessness and social housing organisations such as FEANTSA and CECODHAS, as well as producer lobbies. These actors put pressure on EU institutions to develop further their role in housing policy. Thirdly, EU action is likely to increase in some of the more 'marginal' areas of social policy, including urban policy and action against social exclusion, which have a close affinity to housing. They are 'marginal' compared with 'core' issues such as health, education and social protection. The increasing significance of urban policy is connected with a greater emphasis in the Structural Funds on problems of cities and urban regions, whereas action against poverty and social exclusion may be used to remedy the disadvantage as a balance to the economic efficiency and anti-inflation agenda of the monetary union.

Yet observed against the great social changes (and great inequalities) in the Eastern part of the EU, the scope of these policies (other than economic policy) is modest to say the least. In addition, they are fragmentary and their implementation is slow. They do not imply change in 'the big picture'. Meanwhile the scale of emerging problems in the housing sector and the stubborn passiveness of some of the national governments – including the Estonian government as reported in this chapter – to try to solve them is alarming.

Many hope that the Western part of the EU could persuade the new members to carry out a more responsible social policy including urban and housing dimensions. For instance, in an unusual plea a respected Hungarian housing scholar Iván Tosics

Figure 12.3a Contemporary multi-family house in Tatari Street in Central Tallinn
Source: Sampo Ruoppila, 2004

(2004) proposes that the EU take a more active role in drawing and implementing common housing and urban development policies in order to push the Eastern part of the EU towards socially sustainable urban development. According to Tosics, the results will be all but sustainable if the task is left to the Central and Eastern European countries (CEE) themselves. I am not assured that the EU should step in to make decisions on behalf of the member states in urban issues, but it is easy to agree with Tosics about the necessity to convince the politicians and the policy-makers in the CEE that it is valuable to preserve the European tradition of socially mixed urban areas, open instead of fenced cities, and solidarity in welfare provision

Figure 12.3b   New municipal rental housing in the Lasnamäe housing estate district
Source: Sampo Ruoppila, 2004

Figure 12.3c Single-family housing area under development in Tiskre in the outskirts of Tallinn
Source: Sampo Ruoppila, 1997

Figure 12.3d Contemporary multi-family house in the Pirita garden suburb
Source: Sampo Ruoppila, 2003

including decent housing for all. In contrast, maintaining an indifferent attitude to rising inequalities and residential differentiation processes points towards high segregation in Northern American style. That can still be avoided, but it is a high time to ask what kind of urban development is preferred and what kind of policies should be implemented which would lead that way? In the discussion, let us not forget that – as Pakaslahti and Pochet (2003: 19) put it – the very raison d'être of a European model (as opposed to an Americanisation of Europe) is [the] social side.'

References

Arco, V. (2004), *Eesti kinnisvaraturu ülevaade 2004. aasta I poolaasta* (*Estonian real estate market review, first half of 2004*). Tallinn: Arco Vara real estate company.

Bruns, D. (1993a), *Tallinn. Linnaehituslik kujunemine* (*Tallinn – architectural development*). Tallinn: Valgus.

Bruns, D. (1993b), *Tallinna kujunemine* (*Development of Tallinn*). In: Raam, Villem (üldtoim.) Eesti arhitektuur, osa I Tallinn. Tallinn: Valgus.

Eamets, R. (1999), 'Some insight views into the macroeconomic performance of the Estonian economy in 1989–1997.' *Estonian Social Science Online* 1/1999, 23 pp.

ESA (2002), *2000. aasta rahva ja eluruumide loendus. IV. Haridus. Usk.* Tallinn: Statistikaamet.

Estonian Human Development Report 1997. Tallinn: UNDP.

Kährik, A. (2000), 'Housing privatisation in the transformation of the housing system. The case of Tartu, Estonia.' *Norsk Geografisk Tidsskrift*, 54, pp. 2–11.

Kährik, A., Tiit, E.-M., Kõre, J., Ruoppila, S. (2003), 'Access to housing for vulnerable groups in Estonia.' Praxis Working Paper No 10. Tallinn: Praxis Centre for Policy Studies.

Kinnisvaraekspert (1997), *Eesti kinnisvaraturu ülevaade 1997* (*Estonian real estate market review 1997*). Tallinn: Kinnisvaraekspert real estate company.

Kleinman, M. (2002), The future of European Union social policy and its implications for housing. *Urban Studies*, 39, pp. 341–352.

Kõre, J., Ainsaar, M., Hendrikson, M. (1996), Eluasemepoliitika Eestis 1918–1995 (Housing policy in Estonia 1918–1995). *Akadeemia*, 10, pp. 2133–2163.

Living Conditions (2000). Tallinn: Statistical Office of Estonia.

Loogma, K. (1997), 'Socio-economic stratification in Tallinn and spatial relocation patterns.' In: Åberg, M., Peterson, M. (eds) *Baltic Cities*. Lund: Nordic Academic Press, pp. 168–183.

Mikhalev, V. (2000), 'Inequality and transformation of social structures in transitional economies.' UNU/Wider, *Research for Action* 52.

Miles, M.A., Feulner, E.J., O'Grady, M.A. (2004) *2004 Index of Economic Freedom*. The Heritage Foundation and Wall Street Journal. http://www.heritage.org/research/features/index/countryFiles/English/2004Index.pdf

MKM (2003), *Eesti Elamumajanduse arengukava aastateks 2003–2008* (*Development Plan for the Estonian Housing Sector 2003–2008*). Tallinn: Majandus- ja kommunikatsiooniministeerium.

Omanikele tagastatud majades elavate üürnike probleemid ja nende lahendusteed (1998) (The problems of tenants living in housing under restitution and solutions to these problems). Tallinn: Eesti konjunktuurinstituut.

Org, A. (1989), 'Korteriolud' (Housing conditions). In: Pavelson, M.(ed.) *Tallinna taastootmismehhanism ja arengustrateegia (Tallinn renewal mechanism and development strategy*). Tallinn: Tallinna linnauurimuse instituut.

Pakaslahti, J., Pochet, P. (2003), *The Social Dimension of the Changing European*

Union. Sitra publication series, no. 256. Helsinki: Sitra and Observatoire Social Européen.

Pichler-Milanovich, N. (2001), 'Urban housing markets in Central and Eastern Europe: convergence, divergence or policy 'collapse'.' *European Journal of Housing Policy*, 1, pp. 145–187.

Raitviir, T. (1990), *Linna sisestruktuurid, faktorökoloogiline lähendus* (*Structures of the city, factor-ecological approach*). Tallinn: Tallinna linnauurimuse instituut.

Ruoppila, S. (2002), 'Elamute arendusprojektid: paiknemine ja tingimused Tallinnas' (Location and conditions of residential real estate development in Tallinn). *Maja – Estonian Architectural Review*, 4/2002, pp. 20–25.

Ruoppila, S., and Kährik, A. (2003), 'Socio-economic residential differentiation in post-socialist Tallinn.' *Journal of Housing and the Built Environment*, 18, pp. 49–73.

Ruoppila, S. (2004) 'Processes of residential differentiation in socialist cities.' *European Journal of Spatial Development*, February 2004, Refereed Article No. 9, 24 pp. Http://www.nordregio.se/EJSD/refereed9.pdf.

Ruoppila, S. (forthcoming) 'Housing policy and residential differentiation in post-socialist Tallinn.' *European Journal of Housing Policy.*

Sailer-Fliege, U. (1999), 'Characteristics of post-socialist urban transformation in East Central Europe.' *GeoJournal*, 49, pp. 7–16.

Social Trends 2 (2001), Tallinn: Statistical Office of Estonia.

Sýkora, Ludek (1999), 'Processes of socio-spatial differentiation in post-communist Prague.' *Housing Studies*, 14, pp. 679–701.

Tallinn arvudes 1992 (Tallinn in statistics 1992). Tallinn: Tallinna linnavalitsus.

Tammaru, T. (2001), *Tallinna linnastu rahvastikuprognoos (Demographic forecast of the Tallinn urban region)*. Tartu: Tartu Ülikool ja Harju Maavalitsus.

TLV. 2002. Tallinna elamuehitusprogramm "5000 eluaset Tallinnasse" (*Tallinn's housing construction programme "5000 dwellings to Tallinn"*). Tallinn: Tallinna linnavalitsus.

TLV. 2003. Tallinna arengukava 2003–2009 (*Tallinn development plan 2003–2009*). Tallinn: Tallinna linnavalitsus.

Tosics, I. (2004), 'European urban development: sustainability and the role of housing.' *Journal of Housing and the Built Environment*, 19, pp. 67–90.

Chapter 13

Development and Planning in Latvia, the Riga Region and the City of Riga

Inara Marana

Introduction

This article deals with the current situation and development trends in Latvia, the Riga metro region and the city of Riga, as well as with issues regarding the development of the planning system. Special attention is devoted to the description of the challenges of the demographic, economic and social situation and changes in the urban environment during the period after regaining Latvia's independence.

The article describes the planning system in Latvia which consists of five planning levels: national, regional, district, municipal and local. The planning situation in Riga where various plans – the long-term Urban Development Strategy of the City of Riga, the Riga Urban Master Plan, the Riga Urban Development Program and the Preservation and Development Plan of the Historic Centre of Riga – are under preparation serves as a concrete example. The article also deals with planning issues in the Riga region which is the capital region of Latvia.

Since regaining its independence in 1991, Latvia and its capital Riga have experienced fundamental economic and social changes, influencing the development of Latvia in many ways. In Riga, the changes profoundly affected the urban environment, image and structure. At the same time, changes in the attitudes of society and professionals towards the urban environment and the management of real estate took place, differing considerably from the approaches within the period of the centralized economy.

For the development and management of the land, planning is one of the most important tools. During the last years, planning has received great attention in many Latvian municipalities. Within a short time, planning has once again turned into an important sphere. The legislative and regulatory basis of planning in Latvia is in place and currently being further developed.

What is the demographic, economic and social situation after more than ten years of independence? What planning initiatives are going on in Latvia ? This article briefly describes the main issues of importance in this area.

The Current Situation and Development Trends in the City of Riga and Latvia

Demographic Situation and Development Trends

At the beginning of the year 2003, the population of Latvia was 2.33 million people, while the Riga region was home to 940,000 people or 43.0 per cent, Riga to 739,000 or 31.7 per cent of the national population.

Table 13.1   Population changes in Latvia, the Riga metro region and the city of Riga after regaining independence

	1990 (in 1,000)	1995 (in 1,000)	2000 (in 1,000)	2001 (in 1,000)	2003 (in 1,000)	Decrease in % from 1990–2003
Latvia	2,668	2,500	2,381	2,364	2,331	12.7
Riga metro region *	1,121	1,025	966	957	940	16.2
Riga	909	825	766	757	739	18.8

* Data based on the Statistical Riga Region, which includes the City of Riga, the City of Jurmala and the District of Riga

Source: Central Statistic Bureau of Latvia: Statistical Yearbooks of Latvia 1990; 1995; 2000; 2002; 2003

The decrease of the population from 2,668,000 in 1990 to 2,331,000 in 2003 in Latvia, from 1,121,000 to 940,000 in the Riga metro region and from 909,000 to 739,000 in the city of Riga in the same period was due to the very low birth and high death rates. It resulted also from the fact that households related to the Soviet military left Latvia after the restoration of the country's independence.

Birth rates in Riga are lower than the average for the whole country. Without constant immigration of young people from the countryside and towns, the number of inhabitants would decrease dramatically. This could also result in Riga being populated by mainly people above working age, triggering respective changes in income and expenses for the city budget.

Presently, the city of Riga has reached a level of development where residential concentration in the city centre is replaced by a further extension of the settlement area into the broader agglomeration. Suburban population densities are increasing.

The increase of the number of young immigrants from foreign countries will highly depend on the level of income and living standards in Latvia in comparison to other European Union countries.

Economic Situation and Development Trends

Since 1991, when Latvia regained its independence, the country has experienced fundamental changes in the national economy. Riga's urban and regional structure has been changed by the new economic forces present since 1991. Recessive urbanization, coupled with a sharp decline in industrial production and associated with the closing of industrial enterprises, is the main factor behind the relatively high incidence of urban poverty in Latvia and other post-socialist countries. Large-scale manufacturing collapsed at a much faster rate than small-scale and service sector related firms could emerge to replace them. The results have been increasing unemployment, under-employment and poverty. Nevertheless, the economic situation in Latvia has improved during the last few years. Some branches of the economy – for example services, commerce, tourism, information technology – are developing rapidly.

Currently, the city of Riga accounts for 54.5 per cent of the country's GDP. As the economic base of Riga, the Riga region continues to undergo a radical transformation in which the industrial component is decreasing and replaced by functions like services, commerce, finance, and tourism.

Nevertheless, even today the effects of economic decline are quite pronounced and complex in Riga and elsewhere in Latvia, where industrialization policies had been implemented for a long time. The loss of employment has meant deprivation of means for thousands of citizens in the cities and towns in the region.

Due to the transformation process towards a market economy and due to radical changes in property ownership, employment patterns have diversified. Recently the level of unemployment has been higher than in the early 1990s. It reached 4.7 per cent in the city of Riga and 8.5 per cent in all of Latvia in 2002. In some regions of Latvia, mostly in the Eastern part, the unemployment level was much higher – 25–30 per cent. The changes in the economic structure and the collapse of Latvian enterprises which were unable to compete with Western enterprises are the main reasons for the increase in unemployment. Unemployment is one of the challenges in the economic structure which demands changes in the qualification of inhabitants.

The intensity of traffic has greatly increased in Riga within the last years, mostly because of the radical increase in motorization. Car ownership levels have increased twofold within the last five years. The biggest share of traffic in Riga is transit and goods transport coming in from the metropolitan region. This is due to the constant growth of inhabitants living outside, but working inside the city of Riga. In 2002,

246 vehicles (heavy vehicles, buses, motorcycles, and mopeds) per 1,000 inhabitants were registered in Riga.

Housing Situation and Development Trends

After the declaration of independence, a massive process of privatisation and restitution of public property started. The housing reforms consisted of various measures. During the housing privatisation process, people were entitled to purchase their dwellings using vouchers. During the restitution of houses, previous owners got back their former private houses which had been nationalized during Soviet times. The rent reform altered the payments for rent and communal services, which at the moment are much higher than during Soviet times. In the 1990s, the process of housing privatisation introduced new forms of property such as private rental houses (in restituted houses) and private apartments. At the same time, a smaller part of the housing stock remained municipal and state property. During the privatisation process, no new rental apartments were added to the public housing stock due to low volumes of new construction and a lack of financial resources. At the moment, there are few municipally-owned subsidized apartments and houses for low-income households.

In 2003, Latvia's housing stock consisted of 82 per cent private, 14 per cent municipal, 2 per cent state and 2 per cent cooperative apartments. Housing reforms during the transition period were marked by an emphasis on the privatisation of state and municipal housing, the restructuring and privatisation of the housing industry, the reduction of supply and demand subsidies and a deregulation of the real estate market. Prices for land, materials and labour were liberalized. Restitution, land reform and privatisation were among the most important reforms enacted by the Latvian government thereby providing the necessary foundations for the development of a real estate market. In Latvia and in Riga in particular, housing is developing rapidly. The volume of housing loans has increased considerably in the last four years, mostly due to lower inflation, decreasing interest rates, growing incomes and growing consumer confidence. But at 4 to 8 per cent, the interest rates are still high. Housing markets emerged as a new reality which have profoundly reshaped the existing urban communities.

The transformation from a centrally planned system where housing construction, maintenance and repair were extensively subsidised the state towards a market-based housing system where households are expected to pay the full price of housing services has created a number of problems. This difficult process of adjustments in Latvia is marked by a shortage of affordable housing in urban areas, the deterioration of existing housing in all tenure types, and a lack of adequate investment mechanisms to sustain the quality and vitality of the housing sector.

Between 1990 and 2000, the total areas allocated for housing development in Latvian cities have decreased. In bigger cities, some housing was adapted to other functions, mostly business. The dismantling of provisional accommodations has also been frequent in the last few years. With the drastic reduction of new construction in

1999, the area used up for new residential buildings was approximately 15–20 times smaller than in 1990. Additions to the market did not cover the losses. From 2000 on, housing construction slightly increased. Starting in 2003, construction levels increased even more , especially in the city of Riga. If the envisioned projects are actually constructed by investors and housing developers in the few next years, the number of apartments will increase more rapidly year by year. The most important reasons for that are the development of the mortgage system, and an increase in economic activities and investments. Housing construction has been especially active in the Riga region close to the administrative borders of Riga, creating so-called single-family housing villages.

The average residential space per person in Latvia is 23.4 m^2, and 22.4 m^2 in Riga. Most of Latvia's inhabitants reside in multi-storey houses. The average apartment has 2.3 rooms for living and the average apartment size is 51.0 m^2. In the city of Riga, 5 to 6 per cent of the inhabitants reside in single-family homes. In 2000, there were up to 11,000 households without separate apartments in Riga. Data shows that there is a lack of housing for households with children and low-income households. There is a lack of affordable single family housing, 4 to 5 room apartments and apartments especially for disadvantaged households with special needs. There is also a lack of apartments with appropriate furnishing. Housing in Riga is very close to the level in many Western European cities regarding its standards of accommodation, but it is lagging behind in terms of space.

Social Situation

The most significant aspects of the social transition in Latvia and Riga are associated with labour market adjustments and social differentiation. In response to structural and macroeconomic changes, labour market adjustment has proceeded through growing unemployment and wage differentiation. In the last few years, salaries and household incomes have increased. Nevertheless, incomes are very heterogeneous, and economic polarization is increasing. The Gini index has grown from 0.3 in 1996 to 0.34 in 2002 and has only stabilized within the last years.[1]

In 2002, the average monthly household income was 88 LVL (133 EUR) in urban areas and 105 LVL (159 EUR) in the capital Riga. The survey data also show a polarization of incomes: wealthier households have an income of 190 LVL (288 EUR) per person, while lower-income households earn only 32 LVL (49 EUR).

Food products accounted for 45 per cent of total consumption expenditures in rural areas, for urban households the ratio was 32 per cent. The second largest expenditure is housing. In 2001, it reached an average of 17 per cent, but in Riga and its surrounding region it was higher.

Research data show that 53 per cent of all households regard themselves as neither rich nor poor. Almost one third (30 per cent) of households admit that they

---

1   The Gini index varies from 0 to 1. It is 0 under the preconditions of absolute equality of the income distribution, and it is 1 if the distribution is just the opposite.

are on the verge of poverty while 9 per cent consider themselves poor. Only 7.7 per cent of all households regard their financial situation as good, and only a few – 0.1 per cent of all households – see themselves as wealthy.

Urban Development in Riga

Riga has to cope with the normal problems faced by all cities and regions in most parts of the word like:

- the increased use of the private automobile;
- a decreasing use of public transport;
- environmental problems;
- social inequalities and tensions; and
- an internationally competitive economic climate.

In addition, Riga is faced with the challenge of revitalizing its old industrial and military territories as well as transforming and reintegrating its waterfront areas.

Knowing the history of urban development in Western Europe, it seems inevitable that the processes of suburbanization are also developing also in Latvia's regions. Riga is one of the cities where this process is developing very rapidly – suburban settlement belts of single family homes are being built around the city. As soon as the economy improves both in the Riga metro region and other regions in Latvia, leading to higher per capita incomes, several transformations and further suburbanization processes can be expected. Some of these have already occurred:

- economic polarization – wealthy households are moving to new single family houses, mostly outside the city, close to the city border;
- serious traffic problems – the traffic movement is increasing;
- environmental problems – the loss of open space and attractive landscapes;
- decay or abandonment of central industrial areas;
- decay of large-scale Soviet era housing areas;
- gentrification and commercial conservation of the core residential areas; and
- suburbanization of industry and commerce.

What are the most important questions and expected changes with regard to land use patterns in Riga ?

- Large parts of the city's industrial structure will be changed. Therefore, opportunities must be found for converting industrial land to other uses.
- Growth in incomes is likely to significantly raise the living standard as well as demand for space. This increase could promote the construction of new housing.
- The increased automobile ownership will result in increased opportunities for car-oriented versus pedestrian-oriented shopping. Retail will be re-established in areas closer to where people live.

- Business and financial services are likely to be the fastest growing sectors of the economy in terms of employment growth. The ability to accommodate these new types of business in the historic parts of Riga while preserving the existing form of the built environment is crucial to the economic health of Riga.
- Changing infrastructure demands could result in the need for land-use change. The increase in car ownership will raise the demand for road and parking space. Riga's role as a major airport hub in the Baltic states is also very important for the city's long term economic development.

The above-mentioned issues are only some of the questions discussed during the development of Riga's city planning strategy, and they are important for the development of the city of Riga and for Riga metro region.

Planning as a Development Tool in Latvia

The Latvian territorial planning system is structured in five levels: national, regional, district, municipal as well as local/detailed.[2] At the national level, it is the Spatial Plan which regulates the use of Latvia's territory. It is binding for state institutions and municipalities, but not for private owners. In 2001, the first national territorial planning document entitled *Review of the Use of State Territory* was prepared.

Development plans were to be worked out at the regional level, based on agreements among districts and local municipalities. Five regions for development planning were established.[3] The biggest of them is the Riga region with 1.1 million inhabitants. Currently, four draft versions of spatial plans for the Vidzeme, Latgale, Zemgale and Kurzeme planning regions have already been prepared within the framework of different international programs. The spatial plan for the Riga planning region is in the stage of preparation at the moment. All regions, except the Riga region, carried out their spatial plans with the assistance and financial support from different international programs. The spatial plan of the Riga region is being developed using local professional specialists and experts and local financial resources.

District Development Plans are developed for covering one district administrative territory. There are 26 districts with an average population of 47,000 inhabitants. Seven bigger cities hold independent status and remain outside of this regional planning framework. The District Spatial Plan, which is part of the Development Plan, is binding for the state and municipal institutions, but not for private owners.

---

2   Planning regions differ from the statistical regions in Latvia. This is due to the fact that they have been established on the basis of political agreements between the municipalities.

3   According to the legislation of Latvia, Development Plans have to be prepared at the regional and municipality levels, including a mid-term Development Program for five to seven years and a Territorial/Spatial Development Plan for twelve years.

Municipal Development Plans must be worked out for each city, town and village in its administrative territory. The Spatial Development Plan as a part of the Development Plan is binding for all public institutions and private owners dealing with the development of land. For the time being, most municipalities in Latvia have not yet worked out their Development Plans.

Presently, the territorial planning system in Latvia is characterised by the following features:

- Legislation on territorial/spatial planning and administration has been prepared, but it consists of various laws and regulations which are not sufficiently aligned with each other.
- There are too many levels in planning system.
- Efficient state control on planning has not yet been established.
- Compulsory land taking (eminent domain) is unsuccessful.
- Institutional cooperation dealing with development planning and land control is poor.
- There is a lack of cooperation between planners and politicians regarding the development of inhabited territories.
- People had no possibility to use their political rights for a long time, therefore they have little practice and information on how to realize their democratic rights.

Development and Planning in Riga and the Riga Metro Region

Planning in the City of Riga

In 1955, the first Master Plan (land use plan) of Riga after World War II was adopted. In 1969, the second adopted Master Plan defined preconditions for a strict functional zoning of the city's territory. In 1984, the third Master Plan kept the functional zoning defined in the previous Master Plans' envisaging logical enlargement of these territories in specific places. During Soviet times, the Master Plans were developed by small professional groups without public participation and discussions. They were secret to the public.

Riga's Urban Development Plan, 1995–2005

In 1995, the first post-independence City of Riga Development Plan was adopted. The basic task of the Development Plan was to determine the city's land use and building regulations. The Development Plan contains the policy for further development of the city, based on the principle of sustainable development – environmental, economic and social coordination in the physical development of Riga. The vision of a sustainable urban future was included in the plan.

The Plan designates the permitted use of land within the city limits. The distribution of land uses within Riga's administrative border showed that the capital was planned as a multifunctional city attractive for business, recreation and living.

This plan differed from the earlier plans worked out in Soviet times in several respects:

- the development process – the plan was worked out involving public participation respecting the rights of the inhabitants and the landowners of the city;
- responsibility – legally, the plan is a policy document of the Riga City Council and was discussed and adopted in an open debate;
- transparency – the plan is open to the public.

In this plan, a great attention was paid to the preservation of the natural and cultural heritage. This was radically different from the previous era when the economic development of the city came first both in theory and practice and people had no vote in taking decisions, environmental issues were disregarded, and the cultural heritage was deliberately suppressed. In contrast to previous plans, the Riga Development Plan was developed with the participation of different stakeholders.

Riga's Urban Development Plan, 2006–2018 Under Development

During the last years, different planning documents have been developed in Riga. Due to the quickly changing social and economic situation, the Riga City Council took the decision to prepare a new Development Plan for Riga. New laws concerning urban development have been prepared since the transition and have to be taken into consideration as well. In addition, the previous plan is only binding until the end of 2005. This and other reasons form the background for the decision to work out a new Development Plan.

According to Latvian legislation, the new Urban Development Plan will consist of the following documents:

- a long-term Development Strategy until 2025 – a framework for the Riga Development Plan.
- a land-use plan and zoning by-laws for 2006–2018. This Plan will define the zoning of the territory (property), construction density, etc.
- the Development Program – a mid-term (seven year) document for planning and investment.

The purpose of the creation of this plan was to promote a sustainable and balanced development in the city of Riga, creating an active, vital and contemporary business, trade, national government and recreation centre with an environmentally friendly and comfortable transportation system, while conserving the typical natural values, as well as the cultural and historical heritage.

Further economic growth can promote the suburbanization of industry, trade and services, spatial changes in industrial territories, an increase of housing construction and claims for a higher standard of living, changes in transport infrastructure, etc. All these changes influence the urban model of the city. Riga's Development Plan is a tool for city management and development control.

Riga's Long-term Urban Development Strategy

The first Development Strategy was prepared in 1996, but it has not been approved. After a long discussion, the City of Riga started the preparation of this document. In mid-2004, the first version of the Strategy was prepared and distributed to the Riga City Council Departments, some experts, and others for reviewing. The long-term Development Strategy will be approved until the end of 2005. The Strategy serves as an umbrella for Riga's long-term comprehensive development, including the vision and long-term goals.

The goal of the strategy is to create preconditions for a balanced, sustainable and harmonised development of Riga in a long-term perspective. The objective of the strategy is to improve the implementation and supervision mechanism for the basic planning of the city development and to ensure the conformity of all legislative documents and decisions with the priorities and goals stated in the strategy. The priorities of the strategy are the following:

- the development of a vital and knowledge-based economy;
- the creation of a highly educated and harmonised society and high-quality living environment;
- the development of an environmentally-friendly transportation system and engineering infrastructure; and
- the development of an efficient city management.

Riga's Spatial Plan (Territorial Plan)

The main factors accounting for the development of the new Spatial Plan are the following:

- The legislative base for development and planning in Latvia has changed over the last few years.
- The existing Development Plan for Riga has been designed at a scale of 1:40,000, making adequate land use planning impossible. It is not compatible with the Cadastre map and therefore not useful for improving the taxation system in the city.
- The defined land use plan in the Spatial Plan is not compatible with the goals of real estate, etc.

Latvia is now a member of the European Union, and the development goals of Riga must be achieved within the context of the recognised guidelines, principles and tools for spatial development. The main goal of Riga's Spatial Plan is the following: to promote sustainable and balanced urban development and create an active, vital and modern centre for business, tourism, trade, administration and recreation with a safe, comfortable and environmentally friendly system of transport preserving the characteristic natural values, as well as the cultural and historic heritage.

The main objectives for the preparation of the Spatial Plan are the following:

- to evaluate the current situation, main development trends and opportunities in Riga taking into account the local, regional and national factors;
- to establish the strategic directions and priorities elaborating the proposals for the urban development in the spheres of economy, transport, housing, engineering and infrastructure, as well as social issues;
- to define the use of the real estate according to the development goals of the city;
- to involve as many stakeholder groups as possible during the preparation of the plan, and to prepare new zoning bylaws.

The most important strategic issues for the city of Riga which have been answered in the new Spatial Plan, are the following:

- Will the city of Riga expand its city borders in order to receive additional tax revenues from suburban inhabitants, or will it develop a compact city pattern, developing existing low-density territories of a low density and focusing on multi-storey areas?
- Should the city of Riga undergo centralized or decentralized development, either first developing a strong city centre with an additional role devoted to the development of subcentres, or first focus on developing strong subcentres in the city outskirts while at the same time keeping the important role of the city centre, but with separate functions carried out in subcentres?
- Which economic branches will be the most important bases for a faster economic growth of the city?
- Which will be the most important housing patterns, what will be the basic housing standard in short, medium, and long term?
- How can the city find land use solutions including the use of city territories in combination with transportation, working places and living places that more effectively take into consideration sustainable development principles?
- Which will be the priority in transportation system – public transport or private cars? What changes should be made in the transportation structure in order to free the city centre from traffic?

- How should the deteriorated territories in the city be transferred and reconstructed?
- What should be the development dynamics of public spaces and their parts?
- How should the transportation links with the greater Riga region be formed?

These are only some of the questions which must be dealt with in the new Urban Master Plan for Riga. Besides the strategic issues, this plan is also dealing with economic development, transportation, housing, infrastructure and environment, all from a spatial point of view. The Spatial Plan is binding for institutions and private owners. According to Latvia's legislation, more detailed local plans for different territories and plots must be developed on the basis of the Spatial Plan.

The Preservation and Development Plan of the Historic Centre of Riga

Traditionally, city centres concentrate many different activities within a limited space. Administrative, educational and recreational institutions are situated there, as well as high-density shopping centres, hotels and other service functions. It is essential to strive for functional balance, ensuring a sustainable development of the centre, versifying, changing and supplementing its functions. In the last years, great attention has been paid to the development and planning of Riga's historic city centre.

The historic centre is an internationally renowned cultural heritage site. The historic centre is recognized as of equal value as cities like Venice or Istanbul.

The historic city centre covers quite a small part of Riga's territory – 435 hectares or 1.43 per cent of the whole city territory. At the same time, this territory is the most important part of the city for the business, services, tourism and exclusive housing. It is a multi-functional territory where 39.2 per cent of city's employees work. The volume of business and investments is 32.8 per cent of all business activities in the city. Over the course of eight centuries, Riga's historic centre has developed as a vital centre for culture, education and public institutions of Latvia. The favourable location at the Daugava river and the Baltic Sea Bay has made the historic centre the most economically developed territory of the Riga region, ensuring sustainable and constant preconditions for development. Nowadays, the development of the Riga historic centre promotes national and international accessibility. Public and private transport ensures mobility of labour and inhabitants from the Riga region to the city centre.

The most unique trait of the historic centre is its architecture – wooden houses, Art Nouveau buildings and industrial culture heritage. The following elements of culture and historic heritage deserve protection: the urban structure or fabric consisting of buildings, streets, squares, parks, water resources and other elements. The Riga historic centre comprises three different city scopes: the Old Town, the Ring of Boulevards and the Art Nouveau centre, each of them possessing its own and different urban fabric.

Figure 13.1a The Historic City Centre

Over the last years, activities in the Riga historic centre have increased, and various transformation ideas in the Riga historic centre have been implemented with no respect for the protection of the cultural and historical heritage. The Urban Development Plan adopted in 1995 contained insufficiently detailed ideas for the protection and the development of the city centre. As a consequence, the policy of the municipality concerning the cultural and historic heritage, the protection of green space and the image of the historic city centre was not consistently implemented. This raised dissatisfaction and discussions among the population concerning modern architecture, reconstructed buildings, and the modernised and constructed buildings, as well as the pattern of the development and management of the culture and historical heritage.

Discussions concerning new construction possibilities in the historic city centre have been going on for a long time. At the moment, all serious new construction

Figure 13.1b The Boulevard Ring

proposals must be approved by the Preservation and Development Council of the Historic Centre of Riga. It is for the sake of avoiding mistakes and finding a balance between the cultural heritage buildings and the new buildings.

Over the last years, different problems have come to the fore, for example the increase in traffic or the deterioration of infrastructure. At the same time, the historic centre has been supplemented with various functions characteristic for many contemporary cities without any serious analysis of the related changes in the urban environment (for example, supermarket development, parking places, sport complexes). All this creates the necessity to integrate culture and historical heritage in the current city infrastructure without decreasing the public value of the centre.

In 2001, the State Inspection for Heritage Protection took the initiative to elaborate a Concept for Preservation and Development of the Historic Centre of Riga 'Vision 2020'. It was presented within the framework of the official opening ceremony of the European Heritage Days in September 2001. Due to the positive responses, the Inspection continued its work. As a result, the conception 'Vision 2002/2020' was elaborated in 2002.

In 2000, the Riga City Council took the decision to design the Preservation and Development Plan for the Historic Centre of Riga. The work started in 2000. At the beginning of 2004, the first version of this Plan was prepared. The deadline of the elaboration of the Plan is the end of 2005.

The goal of the Plan is to promote:

- the protection and development of the characteristic cultural heritage;
- the creation of favourable conditions for the cultural and historical urban environment;
- the protection of nature;
- the development of the city centre for active, modern public services, tourism, commerce, recreation, working and living; and
- the development of a safe, environmentally friendly and comfortable transportation system in the city centre for the benefit of the society and the growth of the city.

One of the most important goals in the preparation of the plan was to define the main political base lines which would enable decision makers and specialists to work out a control system, detailed plans, action programs, and to guide and monitor the development and preservation of the city centre.

The plan was worked out with broad public consultations involving various stakeholder groups. A communication plan for the public involvement was prepared. During the planning process, various methods were used: meetings with stakeholders, conferences, seminars, press conferences, information in the mass media, as well as a regularly updated website on the internet. One of the most important steps was the co-operation with a non-governmental institution, a social policy institute organising joint activities over the course of two years during which the interests of the different stakeholder groups were investigated and discussed and the initial version of the Vision of the Historic Centre of Riga prepared.

The public space in the city centre must be attractive and inviting. Public private partnership (PPP) must be considered in the development of the public space in the city centre in order to attract private sector investments for the development of the public space in the vicinities of their property (for example, benches, lanterns, cycle parking lots). Municipal services must take the lead in planning for mutual perfection in order to have a common organization of amenities and a favourable development of the environment. Connections among parts of the city centre and a unified strategy are very important for the development. Each structural part must be developed taking into account its peculiarity, but in a common and linked way. The plan will be finished in a short time, but the most difficult task lies still ahead – how and by which means to implement it efficiently. Improvement in territories like the vibrant historic centre will be very complicated and demand a complex approach.

Planning in the Riga Region

The Riga region is the capital region of Latvia, as well as its main transport, financial, business and industrial centre. The Riga region is a voluntary association of municipalities, being one of Latvia's regions where independent, neighbouring municipalities began the strategic planning process soon after the re-establishment of Latvia's independence. In 1996, the Riga region began a strategic planning process (to develop the Riga Regional Development Strategy until the year 2020). The goal of the strategy was to promote the development of each municipality of the region and to determine the common development goals, priorities and tasks. The Riga region joint planning initiative is a good example for other municipalities of how to cooperate in planning and solve common problems in nature protection, housing, waste management, transportation, water and energy supply and other sectors. This municipal cooperation is voluntary and based on the understanding of the necessity to solve common development problems in the context of broader regional development and linked to the new European Union policy.

The Riga Regional Development Strategy has been prepared and discussed with the municipalities of the region, different ministries, specialists and neighbouring municipalities, as well as the City of Riga. The Riga Regional Development Strategy was adopted at the end of 1999. This development strategy is the basis for decision-making and mutual co-operation between the municipalities forming the region and a precondition for sustainable development. The effectiveness of the strategy depends, of course, on the commitment of the participating municipalities to attain the defined goals.

During the work at the Riga Regional Development Strategy, a common understanding and coordination of interests among the participants regarding the main regional development ideas was reached. Thus, ideas about regional development not only in Latvia, but also on an international scale have been created. Both the City of Riga and the region have participated in many international projects, for example, Baltic Pallete I and Baltic Pallete II where the idea of a common regional development in the Baltic Sea region was discussed. The Riga region proposed some ideas on how to develop transportation, environment, economy, and information technology in the Baltic Sea region.

During the development of the strategy, an open and active dialogue took place among the municipalities of the region, the Ministry of Environmental Protection and Regional Development, the Ministry of Transportation, the Latvian Association of Municipalities, the Latvian Academy of Sciences, the University of Latvia and other institutions. This dialogue has helped to coordinate interests and to come to an agreement on joint regional development goals and the essence of strategy.

In 2003, the Riga Regional Development Council took the decision to prepare the Riga region mid-term Development Program for seven years and the Spatial Development Plan for twelve years. At the moment, these documents are under preparation and discussion.

Conclusions

Latvia, the Riga metro region and the city of Riga are developing fast. Planning as a very important tool has to cope with various activities. The planning system is developing very rapidly in Latvia, but not always consistently. The current preparation of many plans on different levels shows the complicated situation in this area. Latvia has competent planners, valid laws and regulations on planning and an increasing number of existing plans. However, planning and implementation do not always proceed systematically. In the context of a market economy, the management of urban development affecting private land is far from easy in Latvia today. It has started, but it is slow and needs some improvements at the national, regional, municipal and local levels. Improvements in co-operation, joint resources, interest co-ordination and in taking advantage of opportunities are very essential. Nevertheless, some rather good results have already been reached.

References

Central Statistical Bureau of Latvia (2002), *Household Budgeting Survey.*
Central Statistical Bureau of Latvia (1990), *Yearbook of Statistics of Latvia.*
Central Statistical Bureau of Latvia (1995), *Yearbook of Statistics of Latvia.*
Central Statistical Bureau of Latvia (2000), *Yearbook of Statistics of Latvia.*
Central Statistical Bureau of Latvia (2002), *Yearbook of Statistics of Latvia.*
Central Statistical Bureau of Latvia (2003), *Yearbook of Statistics of Latvia.*
Environment and sustainability profile for Riga (2003), Riga.
Ilgvars, F. (2002), *Economic Forces that Have Shaped are Shaping and Will Shape Riga's Urban Structure.* Riga Forum: Modern City. Conference materials. September 06.2002.
Handbook for Analyzing and Documenting the Urban Economic Base. Experience of the Baltic Capital Cities (2002), Washington DC: The Fiscal Decentralization Initiative for Central and Eastern Europe, 1999.
Hietanen, J. (2004), *Housing Market Analysis in Riga.* Helsinki: Helsinki University of Technology.
Marana, I., Treija, S. (2002), *Large Scale Residential Districts in Riga – a Space for Challenges.* CIB – W 69 Housing Sociology Proceeding 'Challenges and Opportunities in Housing: New Concepts, Policies and Initiatives'.
Marana, I. (1997), *Housing and Emerging Problems in Riga. Baltic Cities – Perspectives on Urban and Regional Change in the Baltic Sea Area.* Stockholm: Nordic Academic Press.
Marana, I. (2000), *Problems and Factors of Success for the Implementation Planning Documents in Riga City. Peculiarities of the Strategic Planning in Post Soviet Countries* (St. Petersburg) (In Russian).
Marana, I. (2001), *Sustainable City and Urban Development. Riga City towards Sustainability.* Riga: City of Riga.
Marana, I., Tsenkova, S. (ed.) (2002), *Challenges and Opportunities in Housing: New Concepts, Policies and Initiatives* CIB – W 69 Housing Sociology Proceeding 'Challenges and Opportunities in Housing: New Concepts, Policies and Initiatives'.

Ministry of the Environment of Finland, Ministry of Environmental Protection and Regional development of Latvia, Division of the Regional Policy and Planning of the Ministry of Finance of Latvia (2002), *Spatial Planning and Land Management in Territories in Latvia.*

Process towards Urban Quality. The Preservation and Development in the Historic Centre of Riga (2004), Riga: Report for COST Action C9. European Co-operation in the field of Scientific and Technical Research. Urban Civil Engineering.

Statistics of the registered cars in Latvia (2002), Riga: CSDD.

Tsenkova, S. (2000), *Riga: Housing Policy and Practice. A Framework for reform.* Riga: Riga City Council.

Union of the Baltic Cities, Commission on Urban Planning and Design (2002), *Public Space – Problems and Possibilities.* Riga.

Vision 2002/2020 of the Historic Centre of Riga (2002), http://www.rvc.lv.

Chapter 14

Valletta: Re-inventing the 'City of the Order' for the 21st Century

Conrad Thake

Origins and Historical Context

Valletta as the capital city of Malta has had a long and chequered history. It was conceived as a planned fortified city in the aftermath of the Great Siege of 1565, when the knights of the Order of St John managed against all odds to repel a massive Turkish onslaught to seize Malta (Bradford 1999). The Order of St John was both a military and religious Order. In 1530, eight years after being expelled from Rhodes by the Ottoman Turks, Emperor Charles V offered in fiefdom the Maltese islands together with the fortress of Tripoli on the North African coast (Schermerhorn 1929, Sire 1994). In sharp contrast to the fertile island of Rhodes, the Maltese landscape was barren and unattractive. An early sixteenth century account describes the main island of Malta as 'merely a rock barely covered with more than three or four feet of earth, which was strong and very unfit to grow corn' (Quintin d'Autun 1536/1982).

The knights hardly had time to establish themselves in Malta when in 1565, the Ottoman Turkish leader Sultan Suleiman ordered a 40,000 strong force of Turks and mercenaries to attack the islands. In what came to be known as the Great Siege most of the intense fighting was concentrated around the rudimentary Fort St Elmo, situated at the tip of the Sceberras peninsula. The Turks were only repelled after a protracted siege that had lasted for well over a month and after the invading forces had suffered considerable losses.

The experience of the Great Siege of 1565 had amply demonstrated that the transformation of the Sciberras peninsula into an impregnable fortified outpost was essential for an effective defence strategy for the islands. As soon as the siege was lifted Grand Master La Vallette petitioned the Pope for assistance in providing the services of a military engineer capable of planning a new fortified town on the Sciberras peninsula. The Italian architect, Francesco Laparelli from Cortona was commissioned with the task of drawing up a plan for the new city of Valletta, to be named so in honour of its founder, Jean de la Vallette (Hughes 1969, Hughes 1970, de Giorgio 1985).

The sixteenth century was an age of new towns, but few if any are so well documented as Valletta. Valletta as the new 'City of the Order' was planned and built in the tradition of the Italian Renaissance Ideal Cities and the historic documentation reveals that military considerations by far outweigh aesthetic considerations; the new town situated on a restricted peninsula had to fit its defences (Hughes 1976). Valletta's *raison d'être* was primarily that of a fortified military city serving the needs of the knights of the Order who ruled over the Maltese islands from 1530 to 1798.

Laparelli submitted his plans to the Council of the Order. As the knights deliberated on the feasibility of undertaking the construction of the new city and raising the considerable expenses that this entailed, Laparelli lost no time in mustering workmen and building materials and in outlining the line of fortifications around the peninsula. He was certainly under time pressure to deliver his designs as there was a great fear of another impending attack by the Turks to avenge their previous defeat. A quotation from the Codex Laparelli, which records in detail Laparelli's thoughts and ideas on planning the new city, is clearly indicative of this heightened sense of urgency.

"Give me time and I will give you life" (Francesco Laparelli (1522–1570) – architect, town planner, and military engineer, responsible for the planning of the city of Valletta, 1566–1569; Quotation from the *Codex Laparelli*, Cortona, Italy, cited in Hughes 1978)

After various discussions, the Order finally approved Laparelli's plan and on 28th March 1566, Grand Master La Vallette formally laid down the foundation stone of the new city. Construction works proceeded quickly. An impressive line of fortifications were constructed around the perimeter of the peninsula with squat and pointed bastions on the land-front and at strategic points along the flanks. The construction of the new city entailed the deployment of an appreciable labour force. Many Maltese had perished during the Great Siege of 1565, others had fled to Sicily and would only return to Malta once works on the new city were at an advanced stage. At the peak of the city's construction it is estimated that there were more than eight thousand workers engaged on the construction of the fortifications. As a result of the acute shortage of local workmen, workers were brought over from nearby Sicily and Calabria.

The topography of the Sceberras peninsula was markedly uneven with a pronounced downward slope from the City Gate to Fort St Elmo at the tip of the peninsula. Various attempts were made to level the ground, prior to implementing Laparelli's gridiron pattern of rectilinear building blocks and parallel streets. However, the knights had to abandon the idea of leveling the peninsula as the limited financial resources, shortage of labour, and fear of an imminent Turkish counterattack dictated otherwise.

The rectilinear street plan of the city was laid out by the time that La Vallette died in 1568. When La Vallette's successor Grand Master Pietro del Monte gave the order to transfer their administrative and residential quarters from *Il Borgo* to the new city in 1571 there were very few buildings that were complete. Over the next two decades building works were intensified with the construction of seven auberges or *hotels* that served to accommodate the knights of the seven different *langues* of the Order of St John. The knights of the Order were drawn from different regions of Europe and belonged to a specific *langue* or language corresponding to the country or region that they came from. Thus, for example a knight from Provence would belong to the langue of Provence and would have lodged within the Auberge de Provence. The seven *langues* represented by the late 1560s were those of Aragon, Auvergne, Castille et Leon, France, Germany, Italy and Provence.

In 1582 a visitor to Malta observed that the new capital was now virtually built-up and that there were few vacant sites. By 1590 there were some 4,000 people residing in Valletta. During the seventeenth and eighteenth centuries, the city flourished into a monumental Baroque city in a similar fashion to other European cities such as St Petersburg, Prague and Vienna. The catalysts for growth were various, but there were two principal reasons. The first one was that the population in Malta was at that time mainly concentrated around the harbour towns of Valletta and the Three Cities that were located on the south-east side of the Grand Harbour. As the population increased, so did the demand for housing in the new capital. The second reason was that the knights sought to project Valletta as a monumental urban showpiece by which to impress both upon its subjects and also to other European states the Order's growing status and prestige (Milanes 1988).

By the eighteenth century there was less concern for military considerations and various projects were undertaken with the intention of housing the administration of the local government and encouraging commerce and trade. The city was gradually transformed into a resplendent Baroque city, the epitome of such opulence being attained during the tenure of the Portuguese Grand Master Manuel Pinto de Fonseca (1741–1773) (Thake 1996). During the later part of its rule, the Order's way of conducting its affairs became more concerned with elaborate ceremonial rituals and self-conscious imagery which eventually gave way to a laissez-faire and even decadent lifestyle. The golden era of the knights came to an abrupt end in 1798 with Napoleon's invasion of Malta.

Valletta is perceived primarily as a stringently planned fortified Renaissance town which acquired over time a Mannerist and Baroque over-layering. However, during the British colonial administration from 1801 to 1964, Valletta also benefited from a number of landmarks that subtly tempered the city's urban landscape; a number of gardens designed in the Romantic Neo-Classical style were established along parts of the fortifications; an Anglican cathedral was built with its landmark bell-tower being a distinctive element within Valletta's skyline and also, an imposing Neo-Classical Opera House which was destroyed during the Second World War.

One of the main challenges lies in grafting the dynamic of contemporary form and function onto a city that remains firmly enshrined in the historical contexts and circumstances from which it evolved. After this brief historical exposition it would be relevant to consider the main problems and challenges that the city is facing today.

Key Problems and Challenges Facing Valletta

Physical Constraints

Valletta has severe physical constraints in terms of topography. As a restricted peninsula entirely surrounded by fortifications it cannot expand in terms of land area. Furthermore, Valletta as a World Heritage City with its high concentration of historic buildings and monuments experiences conservation constraints. Several buildings with high architectural and historic value cannot be physically altered and stringent height limitations apply. Also, its urban morphology of a geometric iron-grid cannot be compromised and new public spaces cannot be created.

Declining Resident Population

Historically, the Grand Harbour urban conurbation of Valletta and the Three Cities has always been the most heavily populated area within the island. During the span of the Order's rule in Malta between 1530 and 1798, the population of the Maltese islands had increased from some 20,000 to 100,000 inhabitants. By the time that the Order was expelled from the island, one out of every five Maltese lived in Valletta. By 1861, Valletta and the new suburb of Floriana had a combined population of 30,000 residents accounting for 25 per cent of the total population of Malta. However, at this time Valletta's population had peaked. By 1901, the population of Valletta and Floriana had declined to 15 per cent of the total population (Thake and Hall 1993).

Since the late 1960s there has been a steady decline in the population of Valletta. In 1985, Valletta's population stood at around 9,200, by 1990 it had decreased to 8,300 accounting for a mere 3 per cent of the total population of the Maltese islands. The current population for Valletta today stands at 7, 262 of which almost 30 per cent are over the age of sixty years. This continual exodus of residents leaving Valletta continues unabated and it is alarming: there is a danger that Valletta will no longer be a city where people live.[1]

High Proportion of Vacant Buildings

The steady decline of the resident population has led to a situation whereby it is estimated that there are some 833 vacant buildings which accounts for 23 per cent of all residential units in Valletta. The majority of these vacant buildings are either situated in the socially depressed lower part of Valletta or within sectors along the peripheral areas of the city.

---

1   Valletta case-study in *Regenerating neighbourhoods in partnership – learning from emergent practices*, publication of ENTRUST (Empowering Neighbourhoods through Recourse and Synergies with Trade), pp. 28-29.

Sub-standard Dwellings and Dilapidated Buildings

It is estimated that 75 per cent of all vacant buildings are in a poor physical condition and require rehabilitation. Most of the inhabited residential units within the lower end of Valletta require physical interventions and improvements. A substantial number of residents living in the older parts of the city are not owners of their residences but are tenants with leases dating back several decades and paying nominal rents that were set prior to the Second World War. These anachronistic rent regulations have further contributed to this deterioration in the physical state of the building stock. The tenants usually have very limited income to finance improvement works, on the other hand the owners receiving only peppercorn rents have no incentive to undertake the necessary repair and maintenance works.

Over-commercialisation within the City Centre

The central core of the City based along the first segment of Republic Street or the former Strada Reale is the main commercial and business district of the city. The block bounded by South Street and Old Theatre Street, and Republic Street and Merchants' Street, hosts a high concentration of retail/commercial outlets and private sector offices and banks. The continual commercialization of this area has led to a further depletion of good-quality residential buildings in this sector as there is a very high demand for shops and offices. Very few residential buildings have survived as these have been converted for commercial use.

Inflation of Property Values

All properties within the central business area have in recent years accrued in value as there is a high demand for retail outlets and offices which are strategically located in the city centre. The last twenty years have also witnessed an escalation in property prices in specific strategic locations within the city enjoying panoramic sea-views over the Grand Harbour or Marsamxett Harbour, large properties along St Barbara's bastions and Hastings Gardens are very much sought after. There is also a small but significant class of relatively new comers to the city, mainly young single professionals or non-Maltese persons usually highly-educated and with an arts background who have purchased historic properties with the aim of converting them as their residence. Still the lower part of Valletta and certain peripheral areas remain unattractive in terms of real estate investment. Such properties are situated in socially depressed sectors of the city and are considered unsafe and too run-down.

Issues Relating to Accessibility

Vehicular and pedestrian accessibility are two factors which have a direct bearing on living conditions within the city. During the day, vehicular accessibility within the city is problematic as on-street parking is saturated and the only public car-park is situated near the bus terminal outside the main city gate entrance to Valletta. Furthermore, the city is characterized by sloping streets and steep flights of stairs which are not a particularly endearing experience to the elderly pedestrian. These stairs which give Valletta its distinctive urban character, once incurred the wrath of the English poet Lord Byron who described 'Valletta as the cursed streets of stairs'. Most of the residential buildings are four or five stories high with considerable stairs and when not serviced with an elevator, the inconvenience to the elderly, persons with a disability or even a mother with a young child is very real and a deterrent to taking up residence within the city.

Management of Vehicular Traffic within the City

The high influx of vehicles during the day due to the high number of employees working in Valletta has had a negative impact on the management of vehicles with the city. Parking of vehicles within the city is highly problematic and virtually impossible after 9.00 am. The public transport system does not operate within the inner fabric of the city although the main bus terminus is located just outside the main entrance to the city. Paradoxically, the restrictions in terms of vehicular mobility favour pedestrian circulation within the city particularly in the central core where most of the commercial outlets and offices are located. In the evenings when the offices and shops close, the city becomes virtually deserted and vehicular access and parking are no longer problematic.

Strategic Policy Issues

An intensive rehabilitation programme was initiated in 1987 with the setting up of the governmental agency known as the Valletta Rehabilitation Project (VRP). A number of public buildings – mainly churches, auberges, and monuments – were in dire need of restoration and in this respect the VRP was quite successful in coordinating restoration and maintenance works.

In 2002, the Malta Environment and Planning Authority (MEPA) approved the local plan for the Grand Harbour area.[2] One of the plan's primary objectives is that of securing the economic and social regeneration of the Grand Harbour area with a specific emphasis on the conservation and rehabilitation of buildings to compatible uses. Such conversions have to be both socially relevant and economically viable. The key strategic goals of any effective rehabilitation plan for Valletta can be listed as follows:

- To consolidate and intensify the level of economic activity within the city whilst protecting the existing small-scale and grassroots commercial activities.
- To render the city more attractive in terms of modern-day living and in attracting

---

2   *The Grand Harbour Local Plan*, Consultation Draft, Area Policies, June 1997. Malta: Planning Authority.

Figure 14.1a Valletta – the Fortified City
Source: Pieter Van der Aa, 1708

Figure 14.1b Valletta skyline
Source: Conrad Thake

Figure 14.1c Valletta Stock Exchange
Source: Conrad Thake

new residents to the city. In this respect it is imperative that Valletta should move away from the current residential composition with its high concentration of lower-income social groups physically concentrated in the lower end of the city to a more physically dispersed and heterogeneous resident population.

- To consolidate Valletta as a prime historical and cultural destination for tourists and visitors to the island.
- Valletta as a World Heritage City has to redefine itself for the future in consonance with its rich past. With its considerable architectural legacy, architects and urban planners have to be creative and sensitive in physically re-defining and adapting historic buildings for new uses whilst still in the process respecting the collective memory of the city.

It is important to seek an integrated approach to urban regeneration and to adopt appropriate strategies for implementation. At a time when the central government is facing a major structural deficit problem it is unrealistic to rely on a predominantly state-funded approach. The participation of the private sector is critical although it does not necessarily follow that the objectives of private sector investment within the city and urban regeneration objectives converge. A more realistic approach would be to explore a partnership approach whereby there is a synergy between agencies such as the Valletta Rehabilitation project and the private sector through the creation of a Heritage Trust.

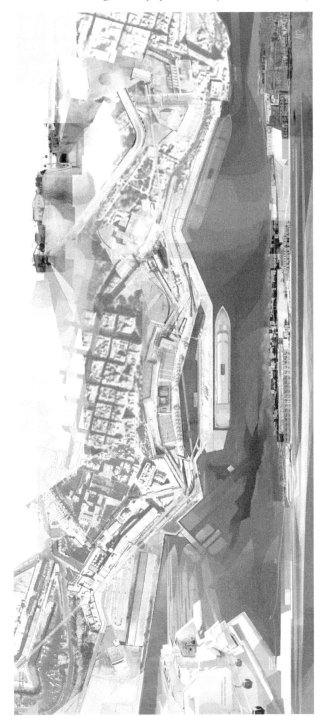

Figure 14.1d Valletta cruise passenger terminal
Source: Architecture Project AP

Case Studies – Reinventing the City

Central Bank of Malta, St James Counterguard, Valletta (1993)

Architect: Richard England (1937–)
Architects once presented with a site usually indulge in the act of building on the site. In this particular instance, the architect had to resist this natural inclination. The site of St. James Counterguard is highly sensitive as it is an integral component of the historic land-front fortifications to Valletta built by the knights during the sixteenth century. Instead, the architect excavated the site and literally inserted the whole mass of the building within the ground behind the fortifications.

Edwin Heathcote in his monograph on the architectural works of Richard England described the Central Bank scheme in the following terms (Heathcote 2002):

> The Central Bank of Malta in the St James Counterguard is another exemplary intervention in that it is commendably self-effacing, a subtle slotting in of a thoroughly modern building into the corner bastion of Valletta's sixteenth century fortifications.

A corridor (which serves as an escape route) runs right around the structure and serves as a boundary between the old and the new while the entrance is through a new rotunda which modulates the building and sets up an axis through its heart. The height of the building is kept below the level of the ramparts throughout (at three floors high) so that the new building is not visible from without, the walls allowing the historic bastion to retain the integrity of its silhouette. The parts of the building are linked by a large central atrium, the form of which is expressed through glass curtain-walls that illuminate the surrounding office spaces.

The strong rooms are, naturally, located in the building's utterly impenetrable basement. The architect compares the task of inserting the new building into the historic walls to the work of a surgeon 'carefully grafting new tissue onto existing skin', and refers to his work as that of the architect performing in the dual role of 'designer of the future and defender of the past.' (cited in Heathcote 2002). This project is a powerful architectural statement based on the synergy of the new with the old.

St James Cavalier Centre for Creativity, Valletta (1997–2000)

Architects: Richard England (1937–)
St James Cavalier is a defensive bulwark that was an integral component of Valletta's land-front fortifications system built by the knights in the late sixteenth century. The purpose of the cavaliers was essentially that of providing raised gun platforms that could be utilized to counteract a land-based attack by an invading army. Externally, the cavalier is an imposing and austere masonry structure with massive blank walls un-punctured by any form of openings or concession to architectural decoration. Internally, the cavalier comprised a large barrel vaulted hall and a parallel series of vaulted chambers; the rest was a huge mass of packed earth fill contained by the

converging outer walls of the cavalier. During the first half of the nineteenth-century, the British Royal Engineers resolved to alleviate the shortage of water supply in Valletta by constructing two circular water reservoirs that were hollowed out of the mass earth-fill of the cavalier.

Richard England was commissioned to transform the cavalier which was in essence a defensive war-machine intended to repel the enemy forces into a modern-day 'Centre for Creativity', attracting as many people as possible from the public domain. The Centre for Creativity was one component of an extensive master plan that encompassed the re-design of the bus terminal, the entrance gate to the city, the site and remains of the former opera house and Freedom Square. The underlying design concept based on the transformation of the cavalier from a war-machine to a peace-time artistic hub is best explained through the architect's own words (cited in Heathcote 2002):

> The object of the rehabilitation of this defensive gun platform was to convert and use its impressive internal spaces as a Centre for Creativity which would become the national arena for both local and international cultural activities. The rehabilitation of St James Cavalier also reflects a pertinent millennium message in the transfiguration of an element of war into an arena of peaceful creativity: the re-use of a weapon of destruction transformed and transfigured into an arena of peaceful creativity. I have always maintained that in order to design what shall be, it is essential, first of all, to understand and protect what has been ... not only by rational analysis but also through sensitive emotional gestures initiated through one's gradual process of *listening to* and *learning from* the stones and spaces of this time-laden edifice.

The problem of converting the internal spaces into meaningful and functional spaces was a challenging one. One of the water cisterns was converted into a theatre-in-the-round (with its underlying steel frame structure being physically slotted within the cavity). The other cistern was fully excavated to form a focal atrium which serves as a congregation space and which is top-lighted by a low lying hemispherical glazed dome. Around the wall of the atrium the architect placed vertical slits through the new walls in the process revealing the fabric of the old excavated reservoir. This design motif was intended to highlight the dialectic between the present and the past, and that material representations of time could be viewed in juxtaposition to one another.

The principle of the different over-layering of historical periods is evident in the parallel series of vaulted chambers whereby the unassuming but elegant vaulted halls constructed by the knights were transformed by the British with the addition of stone arches to create mezzanine levels. Richard England created a new contemporary layer that was diligently introduced without in any manner physically impinging on the original structures. All new interventions were conceived as reversible installations that served a specific purpose. For example, the electrical services and cables were passed through hollow black metal conduits that take the form of freestanding arches independent of the historic building structure. The walls of the long entrance ramp accessible from Castille Square were subtly lit by these illuminated arches of light.

The project was intended to be one precious piece of a larger mosaic intended to revamp the entire entrance area approach to Valletta. Although a detailed master plan was presented by the architect to government this ambitious project has been put on hold mainly due to unavailability of financing.

The Malta Stock Exchange Building, Valletta (1997–2001)

Architects: Architectural design and project management: Architecture Project
Structural engineers: TBA Periti; lighting consultant: Frank Franjou
Climate control: Brian Ford of WSP Environmental Co.

Across the road from the Prime Ministers' Office within the Auberge de Castille, the Malta Stock Exchange operates from a nineteenth century garrison chapel, one of several examples of Neo-Classical architecture in Valletta. The former British Garrison chapel was situated on the very edge of a knights' period bastion. The chapel originally consisted of one large space which was roofed over with a timber-trussed structure, which with the passage of time was in a precarious structural state. The chapel was built in 1855 as a multi-denominational chapel for British naval forces docking along Lascaris wharf, which lies immediately below the foot of the building. Apparently there were numerous debates on where to house this building since Valletta had few vacant building sites. The royal engineer responsible eventually selected the rather precarious site on top of the St. Peter and St. Paul bastion, a site, which lies on top of vaulted casemates and numerous tunnels dating back to the mid-17th century.

The Neo-Classical chapel, prior to the project, was one of many examples of historical buildings that had over time been abandoned. The vision of the Malta Stock Exchange project is a fine example of enlightened patronage – clients who pursued the vision of creating offices that give a new lease of life to an abandoned and dilapidated historic building. The design concept considered the utilization of the whole interior of the former chapel and the direct implication of this decision was the necessity of being able to place a series of 20 metres long steel girders as the main structural components of the modern insertion.[3]

The intervention consists primarily of two arms containing offices running along the entire length of the building terminating with service or circulation towers. These arms are constructed utilizing a visible steel post and lintel structure with glass partitions, whilst open office spaces bridge across below the restored timber trussed roof. The structure abuts onto the original fabric of the building with the use of sheets of tempered glass on both vertical and horizontal planes to create a distinction between the original masonry walls and the inserted structure. In fact, the structure stands completely independent of the historic masonry exterior shell. The steel bracing at the end of each wing is visible and promotes the notion that the

---

3   Project information on the Malta Stock Exchange building was supplied by the architectural firm, Architecture Project (AP). The directors of AP are architects Konrad Buhagiar, David Felice, David Drago and Alberto Miceli Farrugia.

newly inserted structure is entirely reversible and is not an incongruous accretion to the original fabric.

One of the requirements of the Malta Stock Exchange was to ensure that the public may view the pricing of all stocks and shares on display. The public may therefore enter the building and gaze up or down but does not have easy access to the main open office spaces. The voids that characterise the interior also act as buffer zones to limit access to the public. These voids are contoured by sheets of glass acting as railings supported off a metal frame directly bolted onto the 'I' beams that snake their way round the building. The whole structure is revealed, bolts and all, soffits hanging below and railings attached above. The flooring consists of raised tiles allowing services to run below whilst circulation spaces are finished off in maple wood parquet. Clear and frosted glass sliding doors lead into the glass offices, furnished with both maple and aluminium-framed desks. The whole interior is purposely lit in the choice of colours so as to maximize the reflection of natural light that enters the building through the glazed offices on the first floor.

Extensive work was carried out to the original timber roof that was badly in need of restoration. The trusses had been mistakenly loaded (during the latter half of the 19th century) beyond failure point and the whole structure was still upright due to the three dimensional effect of the latticework. Due to neglect, the timber posts were rotting at strategic points and the restoration of the roof entailed the re-modelling of the timber heads with steel shoes, the strengthening of the bottom tensile members with the addition of a steel ties and a complete re-building of the roof structure. The decision to introduce copper cladding followed from studies indicating that copper had been used during the middle half of the nineteenth century but because of its prohibitive cost was then not considered viable. It is now an exciting addition to the unique pitched roof that lies along one of the principal entrances to the city of Valletta, and visible from across Grand Harbour. Apart from the minor changes instituted to the roof there is very little to suggest to the viewer ambling along towards Valletta that the old garrison chapel, now houses six floors of steel and glass.

Valletta Cruise Passenger Terminal Project, Pinto Wharf, Floriana (2001– in progress)

Architects: Architecture Project on behalf of VISET consortium
Valletta's waterfront along the Grand Harbour has for centuries been a thriving hub of mercantile, commercial, and naval activities, both during the later part of the Order of St John's rule when the threat of an Ottoman invasion had abated and during the British colonial period. However, today the Grand Harbour has had to re-invent itself in consonance with the changing political and economic times. Ever since the closure of the British military base in Malta in 1979, the number of military warships anchoring within the Grand Harbour has declined considerably. Instead the Grand Harbour has become a popular port of call for many cruise liners touring the Mediterranean.

The Valletta Cruise Passenger Terminal project, currently underway, is being managed and financed by the Viset consortium. This private sector consortium has undertaken to invest over 14 million Maltese liri over a 4 year period to construct a state-of-the-art cruise passenger terminal along the Pinto wharf together with extensive retail, recreational and cultural facilities.[4] The project is centered around the historic eighteenth-century Pinto Stores, an array of nineteen palatial warehouses built in the Baroque style during the reign of Grand Master Pinto. The imposing façade of Pinto Stores, a part of which was totally destroyed has been faithfully reconstructed as it originally stood and the entire façade has been restored. Restoration has also included the free-standing façade of the adjoining Forni Stores as the interior fabric had been totally destroyed during the war.

The most endearing design concept of the project is the creation of a waterfront experience that would be entirely vehicular-free. With this end in mind, the street that originally separated Pinto Stores from the waterfront was diverted to the back of the stores along the fortifications. Currently, works are in progress to excavate the road and the deep water quay in order to create a water plaza with a pedestrian promenade that would have the restored Pinto Stores as its impressive backdrop. The original sea wall built by the knights would be exposed and would line the boundaries of the new promenade.

The Pinto Stores would have mixed retail/office uses. The ground levels of the stores would be given over to retail outlets, bars and restaurants whereas the middle and top floor levels would be used as offices. The design scheme also includes a sector to be known as 'The Atrium' which would include a crafts centre, an audio-visual presentation, a family entertainment centre and a gaming section.

There are also plans to improve transportation links with the outlying areas beyond the confines of the project area. The consortium are also making a case to have a cable car that would connect the project zone to Blata il-Bajda at the urban periphery, (where a large car-park would be specifically constructed for visitors), and to the Upper Baracca gardens to facilitate direct access to the capital city. Also, the use of water taxis to connect the Valletta waterfront to those at Cottonera and Sliema is also being seriously considered as this would encourage synergy between the different waterfront sectors along the Grand Harbour and Marsamxett. It is anticipated that the cruise passenger terminal, the waterfront plaza, and commercial outlets in Pinto Stores will be fully operational by the end of 2005.

Conclusion

Life in Valletta conjures up imagery of a by-gone age, of some vast museum complex – seemingly restrictive and incompatible with the advent of modernity and yet too historical and stimulating an environment to risk compromising for the sake of

---

4   Information supplied by the VISET consortium. For an overview of the architectural works of Architecture Project refer to 'Mediterranean Mix and Match', chapter 6 of C. Abel, *Architecture, Technology and Process*, Architectural Press, 2004, pp. 203-232.

economic progress. However, freezing the past through an inflexible and naïve sense of nostalgia does not recognize the phenomenon of cities as entities in a continuous state of evolution and re-invention.

Some established local conservation agencies have despite all good intentions vigorously resisted change and any modern architectural interventions within the city. For example, in the late 1980s a project intended to redesign the main entrance approach to Valletta by the internationally renowned architect Renzo Piano was indefinitely put on hold due to widespread criticism from the general public and local conservation groups. The project was deemed to be too disruptive of the historic continuum and incompatible with the traditional image of the city. Mediating between these conservative and progressive forces will determine the physical outlook of the city as a dynamic organism that is responsive to economic, social and political forces.

The challenge of re-inventing a city that what was originally conceived as a military and fortified city during the sixteenth century is a notable one. As Aldo Rossi postulated in his seminal work, *A Scientific Autobiography*, '...form persists and comes to preside over a built work in a world where functions continually become modified; and in form, material is modified'(Rossi 1984).

Today, Valletta is no longer the city of the Order although from an iconographical point of view its architectural imagery is still irretrievably related to it. The city's original *raison d'être* is no longer relevant and Valletta has to re-define and re-invent itself as the capital city of an independent Mediterranean island state that has recently joined the ranks of an integrated Europe. An urban vision for Valletta in the twenty-first century century will have to take into account the country's economic realities and aspirations. Inevitably the city's future will have to be forged upon the solid foundations of the past.

References

Bradford, E. (1999), *Great Siege: Malta, 1565*. Ware: Wordsworth Editions Ltd.

De Giorgio, R. (1985), *A City by an Order*, Malta: Progress Press.

Heathcote, E. (2002), *Richard England*. London: Wiley-Academy.

Hughes, Q. (1976), 'Documents on the Building of Valletta.' In: *Melita Historica*, vol. vii, no. 1, Malta, pp. 1–16.

Hughes, Q. (1978), 'Give me Time and I will Give you Life – Francesco Laparelli and the Building of Valletta, Malta, 1565–1569.' In: *Town Planning Review*, vol. 49, no. 1, pp. 61–74.

Hughes, Q. (1970), 'The Planned City of Valletta.' In: *Atti del XV Congresso di Storia dell'Architettura*, Padova, pp. 305–333.

Hughes, Q. (1969), *Fortress: Architecture and Military History in Malta*. London: Lund Humphries.

Mallia Milanes, V. (1988), 'Valletta 1566–1798: an epitome of Europe.' In: *Annual Report and Financial Statements, Bank of Valletta*, Malta, pp. I–XXXVIII.

Quintin d'Autun, J. (1982), *The Earliest Description of Malta (1536)*. Malta: Malta Interprint Limited, pp. 31–49.

Rossi, A. (1984), A *Scientific Autobiography*. Boston, MA: MIT Press.

Schermerhorn, E. (1929), *Malta of the Knights*. Surrey: Heinemann.

Sire, H.J.A. (1994), *The Knights of Malta*. Yale, Yale University Press.

Thake, C. (1996), 'The Architectural Legacy of Grand Master Pinto.' In: *Treasures of Malta*, vol. II, no. 2, Malta: Fondazzjoni Patrimonju Malti, pp. 39–43.

Thake, C., Hall, B. (1993), 'City Profile – Valletta.' In: *Cities, International Journal of Urban Policy and Planning*, Vol. 10, no. 2, pp. 91–102.

Chapter 15

Problems of Urban Planning in Warsaw

Mareile Walter

Introduction

In February 2004, the planner and architect Adam Kowaleski gave a speech at a conference of the Warsaw section of the Polish Association of Architects (OW SARP 2004), asking: 'Do the present conditions in Warsaw allow for the creation of an ordered urban space?' His judgement on the status quo was that there was 'spatial chaos' in an 'ugly city'. While the overall state of spatial planning in Poland was critical, he said, planning in Warsaw had reached a dramatic crisis.

The conference summarised the results of 15 years of poor spatial planning which had caused irreversible damages to the townscape and the quality of life in Warsaw. There was detailed criticism regarding the following key issues:

- unregulated greenfield development which damaged the former interconnected system of open green spaces and 'fresh air corridors';
- the uncontrolled and uncoordinated increase in building densities and heights and the chaotic development of public spaces and central areas;
- the construction of single-family homes on agricultural or forest land outside the city centre and in the suburban municipalities;
- the severe neglect of the post-war large-scale housing estates.

Within the Polish context, Warsaw is not an exceptional case. Sustainability was not a guiding principle of urban development in post-1990 Poland. In particular, the uncontrolled construction boom transformed the urban landscape. Market forces reigned supreme. Shopping parks, leisure complexes, offices and commercial facilities were built wherever a developer wanted, while the adaptation of the legal and administrative framework for controlling urban development was slow. In the following I will show in more detail how the spatial planning system in Poland has developed since 1990 and what problems emerged. The case of Warsaw and especially the development of retail facilities in Warsaw will illustrate the conflicts between planning needs and societal and political conditions in a post-socialist context.

Planning and Problems of Legitimacy

After decades of a centrally planned economy the term 'planning' acquired a bad reputation in Poland. This applies to spatial planning as well. In many transformation countries, planning is regarded as not compatible with the principles of the free market. In some cases, planning is considered as limiting the development of freedom and democracy. The idea of spatial planning is connected with some of the negative effects of the socialist economy on the urban environment such as the poor social and technical infrastructure, the low quality of flats in the large-scale housing estates or the polluted, now derelict sites of smoke-stack industries in close vicinity to city centres (ECTP 2001). Also, large-scale expropriations based on local development plans led people to think of spatial planning as an instrument of repression (ARL/IGIPK 2001).

Above all, spatial planning suffered from the strictly neo-liberal approach which the post-communist Polish governments took in reforming the economy. The so-called 'shock therapy' of finance minister Balcerowicz consisted of the lifting of price controls, the introduction of private entrepreneurship, the cutting of state subsidies and the privatisation of state-owned companies. As result of these reforms, the Polish GDP fell rapidly and unemployment increased. Large parts of the population sank into poverty, but the national economy recovered in 1992. In the mid-1990s, the country experienced a brief phase of economic boom, entering into a crisis again after the turn of the last century, and recovering again after the accession to the European Union. In January 2005, Poland had an unemployment rate of 19.5 per cent, while the rate of economic growth rose from 1.0 per cent in 2001 to 5.4 per cent in 2004 (Główny urząd statystyczny).

In the atmosphere of change in the early 1990s, any constraint on the 'forces of the market' was regarded as a fallback into the routines of the command economy. Due to this historical burden, neither politicians nor the public supported local development plans or urban development programmes. Controlled spatial development was not highly valued. Therefore it was not a topic on the political agenda and even civic action groups rarely focused on it. The following interview with the then vice-president of the city of Łódź, Wiesław Walczak, in the newspaper Gazeta Wyborcza illustrates the low esteem of spatial planning at this time.

GW: Do you think that another hypermarket is necessary?
Walczak: We live in a market economy. Those who have the money decide if it is profitable for them or not.
GW: But the city is ours. Therefore we want to decide what is happening here.
Walczak: No way. We live in a market economy and you talk like we live in a controlled economy. The city does not decide this, nor does the city president. The people are building the city.
GW: If a company from Warsaw came and wanted to build the Palace of Culture on Piotrowska street, would it be built?

Walczak: Of course, if somebody sells the land and if the building fits onto the lot. It is their money, their plot, and they can build what ever they want. The company won't ask the city president. They alone decide, because it is their money. (Gazeta Wyborcza 2000)

The disregard of planning is accompanied by a lack of financial and personal resources. It is estimated that the administration of the city of Warsaw would need more than 100 additional full-time positions to be able to cope with all its tasks (OW SARP 2004 – Apel Sarp pomaga Warszawie).

Transforming the Planning System

In Poland, the competences for spatial planning are divided between the local, regional and national levels. A distinction is made between tasks of the (central) government and tasks of the (local) self-government. This might seem strange, since all levels are democratically legitimised. However, the historical experiences with the central state powers lead to this distinction.

The introduction of democratic decision-making and the influence of the new market forces caused changes on all levels of the planning system, in particular on the local and regional level. Along with a decentralisation of the system, responsibilities were transferred from the central government to the local authorities. Meanwhile, the process of change still continues: the central administrative bodies attempt to maintain their powerful role, supported by parties like the post-socialist SLD.

The first and most important step in decentralising the state was the reintroduction of independent local authorities in May 1990. Planning thus became the task of the lowest administrative level, the *gmina*. Fairly out of the blue, the communities were in a position to develop their own visions of future spatial, economic, social or cultural development without a given blueprint. As the first bodies that were entirely democratically legitimised, they made important contributions to the construction of the democratic state, yet initially still worked in an institutional framework created by the centralised socialist state (Regulski 2003).

The effort to give maximum responsibility to the lowest possible level led to a break-up of planning authorities. For example, the urban area of Warsaw was divided into several independent communities together forming the Warsaw Communities' Association. Like the communities themselves, this association had a council of its own and the intention of creating an overarching plan for the spatial development of Warsaw. However, it turned out to be impossible to reach a consensus about the development of the central area, which itself had been divided up between several communities. Therefore, the central communities were first joined again into a new community of Central Warsaw. To compensate the local authorities for their loss of power, they were given the status of independent districts with their own councils and their own planning authority (a status specifically created for them). In the course of the administrative reform of 1999, the city of Warsaw was also awarded the status of a county. Warsaw now comprised four independent administrative levels: two parallel ones covering the whole city (the county and the association), the 18 communities of the city and the eight districts of Central Warsaw. The council of the Communities' Association had lost almost all influence on the independent communities, each of which had their own plans and projects which were not necessarily the best choice for the city as a whole.

This confusing structure made coordinated urban development impossible. In October 2002 it was changed again after long struggles with the local bodies. Warsaw became a community with the status of a city and the rights of a county. The city's previous communities became districts. An elected city mayor was given the powers for efficient decision-making and strategic planning. The split of responsibilities between the city and the districts was supposed to be set down in a statute, however this still had not been passed at the time of writing – more than two years after the reform.

In order to enable the communities to become independent bodies, a large share of the state property (in particular buildings, land and technical installations) had been transferred to them in the early 1990s. This provided them with good opportunities to determine the spatial development of the community. On the other hand, with the responsibilities growing and the funds scarce, land was often sold to private developers without consideration of strategic urban planning. Dealing with transnational corporations, the lack of experience and the increased responsibilities meant that local councils were often unable to negotiate outcomes that were favourable for the development of the community. In some places, this led to a form of 'investors' dictatorship' (Tomasz Sławiński in 'Rozwój Budownictwo' 9/2002, S. 42).

Warsaw equally faces the problem that state institutions intend to sell their land. Large recreational green spaces like the central Warsaw *pole Mokotowskie* face real estate development.

Another problem the planning department has arises from the privatisation of its activities (for example, the largest private planning office in Warsaw consists of the outsourced parts of city hall), the very limited number of staff, and the insufficient data on construction projects, property and status of the land. The large transnational developers with their research departments have access to much better data than the local councils (ECTP 2001). The sale of land and the decision on construction applications have also been shown to be open to corruption (in particular large retail complexes, see NIK 2002), which did not improve the reputation of the local councils.

Regionalisation

After the reforms of the local administrations, the reform of the regional and county level followed in 1999. A three-tier system was introduced, comprising the communities at the local level, counties (*powiat*) and regional units (*województwo*). These administrative units do not form a hierarchical relationship but are independent and follow the principle of subsidiarity. The reform was meant to further the reforms of public finance, health and education (Regulski 2003). The 49 old regional units (of the central government) were transformed into 16 new *województwa,* responsible for regional planning and economic development, and taking up some former tasks of the central government (like economic reporting). This is reflected in a dual structure: the president of the *województwo* (the *wojewód*) is appointed by the central government and represents the prime minister, whereas the regional council (*sejmik*) elects the *Marshall* as their representative (ARL/ IGPIK 2001).

This confusing dual character of the *województwo*-level is one of the reasons why the decentralisation of the state in terms of regional development is only a half-hearted undertaking. The other reason is the unwillingness of the central administration to pass on some of their authority and to find new ways of sharing responsibilities with the regional bodies. The lack of access to public finance on the part of the regional and county-levels is a further delaying factor on the way to functioning decentralised structures (Hausner 2001).

An important factor driving the regional reforms is the requirements of the EU for the decentralisation of the state, in order to allow access to structural funds. The new *województwa* are located on the NUTS-II-level, the level where target-1 areas are designated (areas lagging in economic development, below 75 per cent of the EU-average). This meant the creation of an entirely new policy level at the regional level, along with administrations capable of dealing with EU-funding (8.3 billion Euros from structural funds and 4.3 billion Euros from the cohesion fund between 2004 and 2006). Administrations at all levels needed to come up with economic development programmes, and also to be able to monitor and distribute the funds. So far, such experiences were only made at the national level with accession funds from the Phare, ISPA und SAPARD programmes. However, these programmes had rather led to a recentralisation of the structures than to further subsidiarity (Szlachta 2001).

The former *województwo* Warsaw was joined with the adjacent *województwa* into the new *województwo* Masowia, with 5 million inhabitants Poland's largest *województwo.* However, this new administrative unit is also the one with the largest regional disparities: the peripheral areas of Masowia are among the most backward areas of the country, whereas the capital Warsaw is the country's growth pole. With the dissolution of the former *województwo* Warsaw, the region also lacked the administrational body capable of dealing with spatial development in the wider Warsaw region, comprising about 2.4 million inhabitants. Co-operation between local councils is not well developed in Poland, which leads to increased problems related to the suburbanisation of the residential and service sectors. A central conflict revolves around the location of the new motorway bypass around Warsaw, meant to relieve the central city of transit traffic. Here, the new regional planning law of 2003 has enabled Warsaw to develop a plan for the metropolitan area (OW SARP 2004).

Warsaw Case Study

After the political and economic changes of 1989/90 in Poland, Warsaw's economy boomed because of a great inflow of direct foreign investment (FDI). Thirty percent of all FDI coming into Poland went to Warsaw. Banks, insurance companies, financial services and corporations from other sectors (e.g. Daewoo) moved their headquarters to Warsaw. The city provided relatively stable political and economic conditions in order to gain access not only to the Polish market but also to the Baltic countries and the CIS states. It became the financial headquarters of Central and Eastern Europe. The city's advantages over other Polish cities were, besides its status as the capital, a relatively good infrastructure (e.g. telecommunication), its modern industries and the high standard of education of its 1.7 million inhabitants. Helped by the foreign investments, Warsaw became a showcase for the transformation of the economy. Privatisation took off rapidly, employment doubled and the number of businesses tripled (Nowosielska 2000). As a result, Warsaw could further stabilize its leading position in terms of economic power and general wage levels.

As for the urban landscape of Warsaw, the first result of the economic boom was an increase in office space. The location of the new office towers increased the structural problems of the city. The legacies of the past are manifold: Warsaw had been almost totally destroyed during World War II. Most of the historic Old Town was painstakingly rebuilt to its original appearance, and much of the rest covered with prefabricated apartment blocks. The new city centre of the socialist era was developed a few kilometres south-west around the central station and the Palace of Culture and Science (a present to the city from the Soviet Union). To facilitate the rebuilding of the city after the war, the whole area of the city was nationalized. As all real estate was then owned by the authorities, land prices did not matter to the planning of the new city. Because of that the city is very sparsely built, the Palace of Culture is surrounded by a spacious undeveloped area, the streets in the city centre are very wide and mostly clogged by through traffic. An increased utilisation of these free spaces could mean a marked improvement of the city structure.

However, the land was no longer readily available to the authorities. Because it had been expropriated in 1945, much of it was now subject to restitution claims by the original owners or their heirs. Therefore, building activities had to concentrate on an area west of the socialist city centre. So instead of improving the actual centre of the city, the new buildings led to an expansion of the city centre to the west, in spite of the advantages a more condensed urban structure would have offered (Śleszynski 2002). In designing the new office buildings, the surrounding neighbourhood was not taken into consideration. High-rise buildings are located next to small houses, shadows loom over neighbouring residential areas. Some critics have called

Figure 15.1   City centre and districts of Warsaw
Source: Jones Lang Lasalle

this part of the city a 'heap of arbitrarily arranged boxes' (Gazeta Wyborcza 2003). Meanwhile, most new buildings are erected at the periphery. Only 0.925 million m² of the 2.25 million m² of new office space is located in central Warsaw, 75 per cent have been created outside of the city centre, especially along the westbound arterial al. Jerozolimskie.

In the late 1990s, the construction of new office space slowed down as the market flattened out. In 2002, the office vacancy rate surpassed 15 per cent and was even higher in the city centre (Jones Lang Lasalle 2004). With yields in the office market decreasing, investors turned their interest to the retail business.

Spatial Development of the Retail Business in Warsaw

As in most transformation countries, the Polish retail sector underwent more rapid and more radical change than most other sectors. In the late 1980s and the early 1990s, numerous small and micro-retail businesses were newly established, so that the number of retailers in Warsaw jumped from 125,800

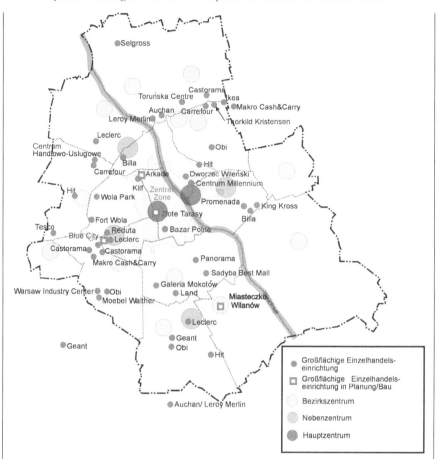

Figure 15.2   Shopping centres in Warsaw (2003) and the Warsaw urban system, based on the 'Spatial Development Plan for the Capital Warsaw' (2001)

in 1988 to more than 450,000 in 1998. Since the mid-1990s however, the involvement of big transnational corporations led to another fundamental change in the structure of the retail business. New retail structures and market concepts were first introduced in Warsaw and later spread to other large cities. Nowadays they have been established even in most small and medium towns throughout Poland.

First, companies like Auchan, Carrefour, Géant and the German Metro established big individual hyper- and supermarkets. Later, shopping complexes consisting of a supermarket as an anchor store and a shopping arcade developed along major traffic arterials in residential areas close to the inner city. The westbound arterial al. Jerozolimskie witnessed a particular concentration of these

**Figure 15.3   Złote Terasy in Warsaw, also showing the Palace of Culture
(top) and the central station (right)**
Source: ING Real Estate Development

developments. In 2000, bigger shopping centres were established in densely
populated areas out of the city centre, like Mokotów and Wola. They also started
to comprise recreational facilities like multiplex cinemas. These malls had retail
areas of 50,000 m², (e. g. the first phase of the 'Galeria Mokotów') and over
100 million Euro of investment volumes (e.g. the 'Wola Park'). In 2004, the two
newly established malls 'Blue City' and 'Arkadia' added another 178,000 m² of
retail space to Warsaw. 'Arkadia', with a retail area of 113,000 m², is currently
the biggest mall in Poland with its own hypermarket Carrefour with a retail area of
18,000 m². At present, more than 50 shopping centres in Warsaw offer a retail area
of about 960,000 m² altogether.

In 2004, the first mall opened in the central city of Warsaw. 'Złote Terasy'
is situated right next to the central station and offers 65.000 m² of retail
space. As in Western Europe, these malls are self-contained spaces with few
connections to the surrounding area. Instead, they themselves simulate urban
life, designed like exclusive downtown areas with streets and plazas. Because
of this, the Warsaw shopping malls also lead to an increase of through traffic
in the central city. Adding to the dull fronts of the buildings, this will lead to
a further deterioration of the quality of life in the public spaces which have
already been neglected for a long time.

These projects are almost exclusively financed by foreign developers or their subsidiaries like the French-American European Retail Enterprises ('Arkadia') or the Dutch ING Real group ('Złote Terasy').

In the 1990s, the development of the big shopping centres was been criticised by small local retailers, by nationalist parties, neighbourhood and environmental groups. The legislators used to welcome the foreign retail companies and the changes to the Polish retail sector they brought with them. The Warsaw administration was proud of the foreign investment in their city. City officials were happy about new shopping facilities for their citizens and did not worry about any possible negative consequences for the urban structure. One member of the city administration even thought it beneficial if people went shopping in the periphery instead of clogging the inner city traffic with their cars.

After the reorganization of the city administration in 2002 (see above), the situation has changed. There is a public debate about the growing number of huge retail areas in the periphery and their possibly negative consequences on the development of the inner city and its public space. The president of the city of Warsaw will not grant permission to any new large-scale project, in spite of applications for more than 1.5 million m² of new retail space (according to the city architect). He now regards the administration as strong enough to influence the development of the market (Gazeta Wyborcza 2004).

With the public awareness growing, the administration getting more effective and corruption being tackled, there is hope that in the future, the urban development of Warsaw will not be influenced solely by market forces any more. Now that Poland has joined the European Union, there are better prospects for a successful urban planning: new funding possibilities arise through the Union's structural funds, and the international discussion of urban planning problems and of communal administration has had an effect on Polish planning practices.

However, the decisions taken in the last 15 years will continue to influence urban development in the coming decades. Because of the size of retail areas outside the city centre for example, it will be hard to find partners for the urgent redevelopment of the area round the Palace of Culture. Nevertheless, the general fear of any kind of planning witnessed in the first years of transformation seems to have been overcome.

References

ARL/IGPIK (2001), *Deutsch-Polnisches Handbuch der Planungsbegriffe* (*German-Polish Handbook of Planning Terms*). Hannover.

ECTP, European Councils of Town Planners (2001), *Spatial Planning in Accession Countries & Implications of EU enlargement for spatial planning.* Proceedings of the Symposium on the 18th of May in Warsaw.

Gazeta Wyborcza, Łódź local issue (2000), 'Znakomity wybór', 15.02.2000, quoted in 'Przegląd prasy lokalnej', www.hipermarket.most.org.pl (28.8.2002).

Gazeta Wyborcza (2004), *'Zapis debaty o centrach handlowych w "Gazecie".'* (*Description of the debate about shopping centres in Gazeta Wyborcza*) (30.03.2004).

Hausner, J. (2001), 'Modele polityki regionalnej w Polsce' ('Models of Regional Policy in Poland'). In: *Studia regionalne i lokalne*, No 1 (5)/2001, pp. 5–24.

OW SARP, Oddział Warszawski Stowarzyszenia Architektów Polski (2004), Conference 'Wpływ nowych uwarunkowan ustrojowych i prawnych na jakość i kształtowanie architektury oraz ładu przestrzennego w Warszawie' ('The influences of new administrative and legal conditions on the quality and form of architecture as well as spatial structure in Warsaw'), Warsaw, 19.2.2004.

NIK, Najwyższa Izba Kontroli (2002), 'Informacja o wynikach lokalizacji dużych objektów handlowych (super i hipermarketów)' ('Supreme Chamber of Control: Information about the results of the control of the localisation of large retail complexes (Super- and Hypermarkets)'). Białystok.

Nowosielska, E. (2002), 'Sektor usług w aglomeracji Warszawskiej w Latach Dziewięcdziesiątych' ('The service sector in greater Warsaw in the 1990s'). In: Węclawowicz, Grzegorz (ed.), *Warszawa jako przedmiot badań w geografii społeczno-ekonomicznej.* Warsaw, pp. 195–228.

Regulski, J. (2003), *Local Government Reform in Poland. An Insider's Story.* Local Government and Public Service Reform Initiative, Open Society Institute, Budapest.

Śleszynski, P. (2002), 'Delimitacja centrum Warszawy – problemy badawcze' ('Scientific problems of delimiting the centre of Warsaw'). In: Węclawowicz, G. (ed.), *Warszawa jako przedmiot badań w geografii społeczno-ekonomicznej.* Warsaw, pp. 65–102.

Suchowski, K. (1999), *Finanse samorządu terytoryalnego w teorii i praktyce* (*Finances of territorial self-government in theory and practics*). Kraków.

Szlachta, J. (2001), 'Polityka regionalna Polski w perspektywie integracji z Unią Europejska' ('The regional policy of Poland from the perspective of the EU-accession'). In: *Studia regionalne i lokalne*, No. 1 (5)/2001, pp. 25–40.

Walter, M. (2003), *Einzelhandelsentwicklung in Polen. Probleme der planerischen Steuerung des Einzelhandels in Warschau.* (The development of the retail sector in Poland. Problems of planning the retail trade in Warsaw). Unpublished diploma thesis, Institute for Urban and Regional Planning, TU Berlin, May 2003.

Chapter 16

Central Europe's Brownfields: Catalysing a Planning Response in the Czech Republic

Yaakov Garb and Jiřina Jackson

Introduction

A brownfield is developed land (such as a factory site, railroad siding, or former military base) that is now underused – often vacant or derelict, and sometimes contaminated or feared contaminated. While the market will usually 'recycle' land whose former use has become obsolete in some way, brownfield properties are 'stuck' in an underused state. The market, left to itself, will not recycle them into more active use, often because the perceived cost and risk of bringing them back into use exceeds the benefits to owners. The broader urban and social costs of these underused 'holes' in the urban fabric are great, and would often justify the necessary expenditures. Land in central and accessible locations lies non-productive, in terms of the urban fabric and municipal revenues, while marring the contiguity and desirability of adjacent properties. To eliminate the barriers and market failures that prevents such land from reentering into productive use, some kind of public sector intervention (finance, coordination, regulatory change) is often necessary.

This chapter describes how the urban planning issues surrounding the problem of brownfields play out in the Central European context, and the role of a non-profit advocacy organisation in generating change. While the issue is well recognised in Western Europe (as well as the U.S.), and a significant professional literature on the topic has developed, the particular circumstances of Central Europe produced a unique kind and scale of brownfield problem, and posed regionally specific planning and institutional barriers to their solution. We describe these particular circumstances and challenges, and why this kind of cross-cutting and complex issue was particularly amenable to NGO interventions. In particular, we describe the catalytic work of one NGO in raising awareness of the brownfield issue, and helping lower barriers to their solution. The interaction of the brownfield problem with EU accession is also discussed, including some suggestions for making EU funding and planning categories more capable of facilitating brownfield reuse.

Our goals are to illustrate how one pressing urban problem and its remedy play out in the Central European context; to highlight the many levels at which change must take place in order to effect change on a complex (multi-disciplinary and multi stakeholder) planning problem in a transitional economy; to use the experience of one organization's advocacy to illustrate how the non-governmental sector can effect systemic change on this kind of complex issue; and to point to the promises and deficits thus far of the EU accession process in assisting in these solutions. Most of our examples are drawn from the Czech Republic, but apply with some variation throughout the region.

Extensive Brownfields as a Post-Socialist Legacy

Brownfields are produced as a regular outcome of industrial restructuring in any country. But the dynamics of socialism and the circumstances of transition from a socialist to a market economy, both described below, left Central European countries with an exceptionally large burden of brownfields, especially urban brownfields, and with a greatly lessened ability of the market to 'recycle' these properties into productive use.

The processes by which rural and exurban brownfields in Central Europe developed would be familiar from Western European and U.S. contexts. Historically, industries were located close to their energy sources. Thus, glass manufacturers were established in mountains areas, close to sources of timber, and changed locations when their fuel areas had been lumbered. Textile industries were situated adjacent to the streams that powered them. And steel, chemical and other coal-dependent industries were located close to sources of coal, on sites that were abandoned once coal extraction ceased after nearby sources were depleted or when mining became uneconomical.

These closures and relocations of production facilities often left polluted and dilapidated sites, and had drastic social consequences. In Poland, in Silesia, the former centre of the mining, steel and chemical industries, large areas are utterly contaminated. The areas of former steelworks in Ostrava (Czech Republic) are not much better, as are large areas abandoned by the socialist chemical industry. An atlas of industrial production in the socialist Central Europe would nowadays also serve as a fairly good map of the contamination of buildings, soil and water in the region. In addition to the larger and now notorious facilities in such an atlas are thousands of smaller operations all over the Central and Eastern European Countries (CEEC), which are now seriously contaminated, and left in a dilapidated state in the aftermath of industrial and political transformations.

An exceptional feature of Central European brownfields compared to Western Europe, however, is their abundance in *urban* settings. This is the result of the shared socialist heritage of CEC cities. Under socialism, with no real estate or capital markets to speak of, state companies did not consider the cost of land or of money when making construction or operating decisions. Sometimes, ideological or political considerations dictated location: such as in the location of a large steel industry next to Krakow, in order to dilute the notorious intellectual and religious character of the

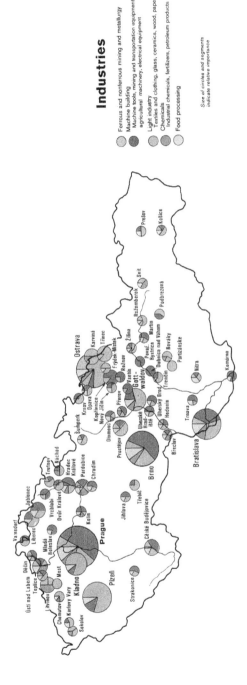

Figure 16.1   Czechoslovakia Industries, 1974

Source: http://www.lib.utexas.edu/maps/czech_republic.html. Map No. 501820

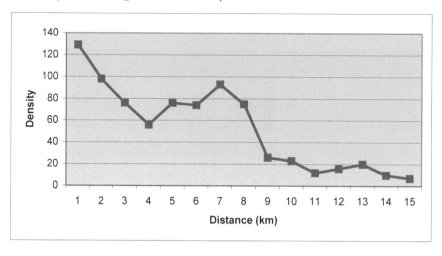

Figure 16.2   Densities in built up areas as a function of distance from city centre: The density 'hump' typical of CEC cities

Data Source: J. Brzeski, 'Guidelines for Developing Local Land Management Strategy,' Krakow Real Estate Research Institute, 2001

city. Thus, production facilities were situated in what would otherwise have been prime urban sites in or adjacent to central locations.

Additionally, in command economies, long range plans and quotas regulated raw goods allocation and production. The inflexibility and bad predictions about demand and supply associated with these spurred production facilities to set aside large areas for the storage of raw materials and finished products, often for extended periods. In the absence of market costs, these facilities were relatively insensitive to the spatial and financial inefficiencies entailed by these build-ups, and their premises were often much larger than their counterparts in capitalist economies.

Thus, post-socialist cities have a legacy of comparatively more and larger industrial sites in cities, as illustrated in figure 16.1. Central European cities (even those that are not heavily industrial), have 2 to 3 times the amount of space devoted to current or past industrial uses than their western counterparts. The portion of land devoted to industrial uses is even higher in distinctively industrial cities, and these face massive brownfield and restructuring problems with the demise of their indigenous industries.

Another aspect of socialist planning further added to the formation of large industrial sites on what is today quite central and valuable land in Central European cities. Massive high-rise housing estates were built beyond city edge industrial sites (often to house workers in these industries), enveloping these industrial sites within the city. Because of these densely populated housing estates, the usual curve of declining density as one moves away from city centres is interrupted by a large 'hump' in central European cities (see Figure 16.2).

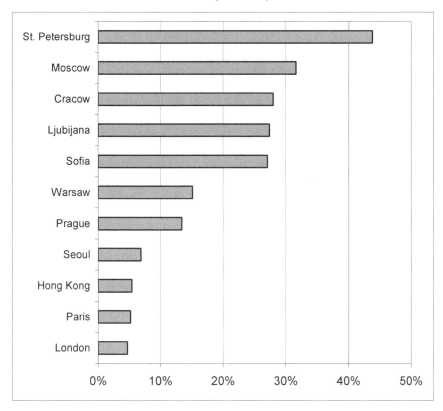

Figure 16.3   Percentage of Urban Land Devoted to Current or Past Industrial Uses

Source: Alain Bertaud, 'The Spatial Structures of Central and Eastern European Cities: More European Than Socialist?', paper presented at the 'Winds of Societal Change: Remaking Post-Communist Cities,' University of Illinois, June 18–19, 2004

With the change in regime in the late eighties, much CEEC industry found itself rapidly redundant, unable to compete in broadened markets terms of its efficiency and the products it offered. A spiral of decline commenced. Maintenance (already a weak point in socialist economies) virtually stopped with the change in regime. Cash-starved companies sold their production and maintenance equipment. The functioning of companies sold in the privatisation process was often short lived, only deepening their dilapidation of their property holdings. Indeed, sometimes these sales were not intended to yield a functioning firm, but were for the purpose of asset stripping. Some properties were rented in a dilapidated form, and on an oversupplied market, and put to secondary uses (for car breakers, for example) that could further contaminate the land.

The physical degradation of these sites was accompanied by a degradation of their ownership status and integrity. Through the privatisation process, and because owners and bankruptcy administrators tended to dispose of properties in small individual lots, these often became less viable purchases for redevelopment. Other sites were unsellable because they were securities for mortgages, often valued at hundreds of percent their actual value.

In addition to the brownfields caused by industrial obsolescence, additional brownfields arose through demilitarisation that emptied large army bases in and around CEC cities. Extensive railway lands and siding areas, often quite polluted, are also drastically underused in many Central European cities.

The extent of Central European brownfields did not appear immediately with the transition, but grew as firms failed and properties degraded. Brownfields began to be a policy issue during the privatisation process, and especially with the visible reluctance of foreign investors to purchase sites that were or might be contaminated (combined with their eagerness for greenfield sites at the city edge.) With more and more untouchable sites dotting the urban landscape, the scale and seriousness of the brownfield problem in Central Europe became visible.

The urban consequences of the patches of underused land within cities are intensified by the sprawling patchy development of greenfields for commerce, industry and housing outside cities. Shrinking urban populations, and the competition from the easy exurban development on greenfield sites make it even more difficult to bring brownfield sites back into productive use.

The Barriers to Brownfield Reuse: The Czech Example

Unlike many urban problems, the question of brownfields is relatively conflict-free, in that its solution is not, fundamentally or substantially, to the benefit of some stakeholders at the expense of others. More than most issues, the recycling of urban land, is a win-win solution, benefiting a range of stakeholders in the private and public sectors, at various scales.

Still, it is a obstinate problem. The difficulties lie elsewhere: in the complexity of the issue and of the linkages and coordination required for a solution (linkages across several disciplines and a multitude of stakeholders, for example); in the absence of a clear-cut locus of responsibility; and in the diffuseness of the benefits to be gained from an overall solution, whose appreciation requires a fairly sophisticated understanding of urban dynamics.

At a very basic level, the very absence of the recognition that brownfields constitute a category is a problem. It was not clear, initially, that a dilapidated railway siding here, an abandoned Soviet army barracks there, and a contaminated factory in a third place, all constituted a single kind of issue, indeed a serious one. A further problem was the lack of any data and measures on how extensive brownfields were: no clear-cut definitions of underused land, nor of the various kinds and parameters of such land, no registry of sites, nor estimate or mapping of their extent, were available.

And, even with the dawning recognition of brownfields as a problem with a name, there was still confusion and fragmentation regarding the locus of responsibility, leadership, and coordination for mobilizing around the issue. For example, in the Czech Republic, these resided among a multiplicity of agencies. Thus, it was the *National Property Fund* (a state agency responsible for the privatisation process), which instituted Environmental Clearance Contracts to reassure prospective purchasers in the second round of privatisation regarding the sometimes crippling liabilities associated with potentially and actually contaminated state-owned properties privatised in the 'first wave' of privatisation after the political transition in the late eighties. The *Ministry of Environment*, was a useful technical consultant and supervisor for site cleanup, and a first point of recourse for sites posing a substantial environmental risk. *CzechInvest*, the government investment promotion agency under the responsibility of the Ministry of Trade and Industry, charged with developing and promoting sites, primarily for foreign investors, was prompted to develop brownfield awareness and skills by the recurring questions of potential investors regarding urban sites. The *Ministry of Regional Development*, traditionally charged with the formation of national level planning, with the re-formation of a regional level of administration, and with links to and training of local administration, was well positioned to be concerned with a national strategy on brownfields, or with the training of local authorities to inventory and assess their brownfield holdings. The *Ministry of Trade and Industry* itself, which had long been responsible for the environmental rehabilitation of depleted mineral extraction sites, was also a repository of important experience and initiative. And because many *Ministry of Finance* departments touched on issues that related to brownfields, because it often exerted considerable coordinating capacity, and because of its ability to strategically recognise and act on the long-term costs of deferred problems, this Ministry might also be a key locus for taking leadership on the brownfields issue. Yet, despite, and perhaps because of, the multiplicity of relevant authorities, no clear-cut leader on the brownfield issue emerged, which would coordinate the activities of these and other players, and set a tempo for their engagement with the issue.

Associated with the absence of a locus of leadership and coordination, was the absence of a strategy and guiding principles. Key choices needed to be made: by whom should the thousands of sites be counted and assessed, according to what principles should their treatment be prioritised, where should the funds for this come from, and what were the goals and targets that would drive this effort? Without a clear-cut 'owner' of the problem, and a clear strategy for tackling it, it would be difficult to galvanise the necessary political will to make changes across a range of institutions and spheres.

Besides these abstract problems of (lack of recognition, institutional locus of responsibility, and strategy), a set of more mundane impediments and a lack of tools, hampered the market uptake of brownfields.

Some of these impediments might seem accidental and obscure, yet for a developer pursuing a brownfield site, the devil really was in the details of administrative rules and procedures. For example, with many large single sites broken into multiple ownership through the privatisation process, complex land assembly was often a prerequisite for a large project; yet the legal instruments for this were quite uncertain.

Months and years of work could go down the drain with a single recalcitrant owner. Something as small as the 30 year depreciation terms for the costs of demolition and environmental clearance, made brownfield sites a relatively unattractive investment, compared with other activities, whose costs could be depreciated over a far shorter period. The local planning system was often inflexible and arcane, requiring, for example, much learning and frustration in order to change the zoning status of a former industrial site into a more appropriate commercial or residential use. The levels of cleanup demanded of contaminated sites were sometimes overly uniform – not sufficiently discriminating of the intended end use of the site, whose ordinary ambient state would be far below the levels of cleanup demanded for 'rehabilitation.'

It is no wonder, then, that, in the first decade or so after the transition, playing by the rules in obtaining, rezoning, clearing, and redeveloping brownfield sites was too frustrating for *bone fide* developers. As they abandoned attempts to redevelop brownfields, the field was initially largely left to the less careful or informed. Local planers and environmental administrators, as well as local property buyers and stakeholders often did not realise all the risks and complexities entailed by urbanised land reuse. This led to the development of a niche for developers who cut corners, though in doing so they were able to develop brownfield properties only to the lower standards of a local market, since such crudely redeveloped properties were unlikely to pass review by the competent due diligence teams employed by large international investors. This was compounded by the fact that certain permits could be purchased, rather than legitimately obtained, and professional consultants could be found who would look the other way.

A more serious impediment to brownfield rehabilitation was the structural disincentive posed by the ready availability of greenfield sites. In essence, by providing connecting infrastructure (roads, sewage, electricity), and ignoring the externalities of ex-urban uses, the government was subsidizing the development of greenfield sites, and undermining the relative appeal of brownfield sites despite their more spatially efficient locations. Recycled brownfield sites in most major Central European cities are, in principle, sufficient to support years, if not decades, of new development, which is instead leaking out to the office parks and hypermarkets at ribbon sprawl locations on radial highways.

A final and important kind of barrier to brownfield reuse is the lack of technical tools and professional know-how. Some of these are quite simple: such as a simple method for local authorities to audit and prioritise their brownfield holdings, or a nation-wide registry of contaminated sites and their parameters, linked to the cadastral registry, so that buyers can unambiguously know the status of their sites, and sellers can record the kind of cleanup performed. Similarly, a compilation of the unit costs of various kinds of clearance and cleanup procedures would allow a better ability to foresee and benchmark a proposed project, and prevent unscrupulous prices for work performed. More sophisticated financial, legal, and administrative tools are required (and were lacking), for example, to support the kind of public-private partnership arrangements, which would be necessary in order to mobilise around and raise capital for brownfield projects.

The Role of an Advocacy Organization in Addressing a Complex National-Level Problem

The previous sections have described the massive extent of brownfields problems in post-Communist cities, and the substantial barriers to tackling these. Despite talk of the importance of urban land effective management, somehow, amidst the pressing issues of transition and the priorities that were a condition for EU accession, these land use issues did not receive the same kind of national attention or outside technical assistance. Though the questions of urban land availability and land recycling are clearly linked to broader issues of structural change and competitiveness, EU subsidiarity principles placed land use and planning issues out of the bounds or at least the central thrust of EU involvement. The 6,500 or so Czech communities, which have primary planning powers, might have been a key player in addressing the brownfield sites that so many of them possessed; but, as we have seen, they lacked the know-how and overview necessary to mobilise necessary national level changes. These circumstances created a leadership vacuum, that made NGO interventions of special importance and leverage. This chapter describes, through the experiences of one NGO, the kind of networked advocacy that was attempted, and what it was able to achieve. . Through this, we hope to lay out the kind of catalytic potential the NGO niche offers in facilitating comprehensive change in a planning regime.

Several NGOs began work on the brownfield issue, some early on, and others have joined in more recently. For example, the Vankovka civic association in Brno was active consistently over a long period in lobbying for the rehabilitation of the old Vankovka factory site, close to the centre of town, so that this could be developed into a thriving commercial space with areas devoted to community use. The brownfield project of The Institute for Transport and Development Policy (ITDP), whose brownfield work phased into the work of the Prague-based NGO ('civic association'), IURS, was one of the early leaders in the brownfield advocacy area, and its work, in which the authors were involved, is described here. The Prague-based Institute for Environmental Policy also took up brownfields issues, preparing an analysis of the legal framework for brownfields rehabilitation. More recently, one of the strongest Czech NGOs, the CTKP (Centre for Communal Programs), has begun to focus on the brownfield issue

The NGO described operated out of Prague, with a three-year mandate and funding to work on 'smart growth' issues (sprawl restrain and city revitalization) in Central Europe.[1] This organization, with in-house skills in both the property

---

1   The Institute for Transport and Development Policy (www.itdp.org), based in New York, and operating internationally, had been operating on transport issues in Central Europe since the early 1990s. A grant from the Rockefeller Brothers Foundation for 'smart growth' advocacy enabled the work described in this chapter. IURS, a Prague-based civic association

development cycle and in public interest advocacy, had a considerable advantage in the situation described above, of a serious problem, largely unidentified, demanding new insights and tools, and coordinated action across sectors and disciplines. Various freedoms allowed the organization to be responsive to this emerging issue, and focus on mobilizing brownfields at a national and regional scale. Not being bound to a particular institutional or disciplinary agenda (as various government agencies are) nor geographic scale (as municipal and local actors are), the organization was able to consult with a variety of stakeholders and build a systemic viewpoint. The luxury of multi-year funding, and the knowledge that it was to facilitate change in a field, rather than 'set up shop,' allowed the organization to focus on and do what was best for the topic at hand, rather than being driven by PR considerations. Thus, as opposed to a consultancy or even a government ministry, the NGO was able to 'give away' away ideas and expertise, rather than hoarding them for repeat sale or building institutional power. The NGO 'do-gooder' status made the organization less threatening, initially, to other stakeholders, who could collaborate and discuss difficulties more freely (though some players became more nervous, over time, as the NGO began to generate analyses and recommendations appeared to highlight their own inaction).

Because an advocacy NGO's resources are nimbleness and independence, rather than large resources or institutional power, it must create change catalytically, rather than by brute force. For example, it can do things such as

- educating and embarrassing other larger players into action;
- networking players with one another;
- importing knowledge and best practices;
- demonstrating in pilot projects what might be done more broadly;
- generating and exhorting a strategic viewpoint or plan;
- cultivating and nourishing individuals in key locations within formal institutions; and
- quickly creating small but pivotal tools or pieces of knowledge, that can leverage larger change.

In this case, the NGO began its work with a kick-off conference on the topic of brownfields, and invited a range of panellists (from various ministries, local and regional authorities, the private sector, and other NGOs) to give a short presentation structured by a set of questions regarding their own organization's or sector's response to 'the brownfield issue,' basic background

---

devoted to urban sustainability issues, was established by the project, as partner and eventually an inheritor of the Central European initiatives started under this 'smart growth' mandate (see www.brownfields.cz). Jirina Jackson, based in Prague, acted as the project's Central Europe coordinator, and conducted most of the work on the ground, in collaboration with Yaakov Garb, ITDP's Director of Central European programs.

on which was provided in conference preparation packet. Once a couple of key ministerial players had agreed to talk, other ministries could not afford to not be absent, even if they had little or no idea about the topic. This sent a wave of (sometimes frantic) preparation through several organizations, in order to prepare a description of 'what we are doing about the brownfields crisis.' In several cases, the hosting NGO itself was called on to do a good deal of the coaching on the preparation, providing important opportunities for mutual education. High-profile sponsorship for the workshop by the city of Prague was obtained, which had the happy side-effect of requiring formal approval by the City Council, whose members then needed to review a background briefing on the purpose of the workshop before voting. In short, much of the value of the workshop was achieved before it opened. The workshop itself generated a baseline mapping of what the various stakeholders knew and were doing regarding the topic, the barriers they perceived, and an initial body of expert and local knowledge that could be distributed. The latter benefited from overseas organizations (OECD, the U.S. Environmental Protection Agency, and others), which contributed experts, and materials for translation – the first such materials available in Czech, which was especially important in reaching a more senior generation of civil servants and decision-makers who did not speak Western European languages.

Thus, this workshop launched the issue of brownfields visibly on the public agenda, mapped the issue and the key barriers to brownfield reuse, distributed an available body of knowledge in Czech and English, and created a cross-disciplinary and cross-institutional network of people with a familiarity and some degree of concern for the issue.

A second conference, which attempted to address specific questions, such as financial and insurance arrangements for brownfields, or public participation in restoration, turned out to be premature, and somewhat less successful, as the 'market' was not ready for these specialised topics. Instead, the NGO began initiatives that worked more closely with ministries, regional and local authorities. A series of brownfield presentations were given to regional authorities, and to towns with considerable brownfield holdings. This educated them about the issue, and how they might inventory their brownfield holdings, prioritise sites for urgent treatment, and conduct preliminary planning exercises for these sites. [See the box Šternberk case study] It also increased the NGO's knowledge about the barriers experienced at the local and regional levels. Over time, this awareness at local and regional levels began to percolate up to become pressure for the necessary administrative and legislative reforms at the national level.

Along the way, the NGO identified and created some specific high leverage tools and pieces of knowledge. A quick rough estimate of Prague's brownfields showed one thousand hectares of unused land in central areas: a galvanizing figure that helped prompt a more systematic survey (see Figure 16.4).

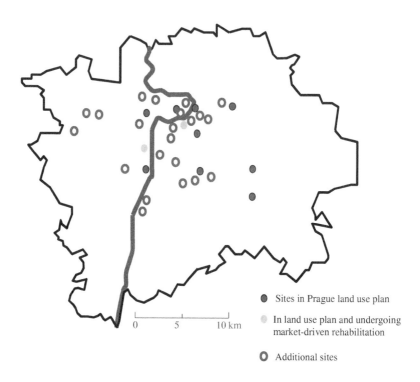

Figure 16.4 **An early schematic map, used for lobbying, of the location of major brownfields in Prague**

Based on its work with municipal authorities, the NGO crafted and circulated a 'brownfield inventory' methodology, which towns could use to assess and prioritise their holdings. It also began circulating information on how various EU funding categories might be applied toward brownfields projects, and helped cultivate several grant applications, some of which were subsequently successful.

The various government ministries must have a large role in any comprehensive brownfield solution, and the NGO worked quite intensively with some, supporting initiatives and capacities within the ministries, while pushing to deepen inter-ministerial linkages on the subject. Thus, the organization wrote a position paper that was adopted as the ministerial strategy paper for one of the ministries, helped encourage another to establish a cross-ministerial brownfields committee, under the direct responsibility of a Deputy Minister, and encouraged (and helped draft the terms of reference for) external grant categories for brownfields work in two other ministries. It helped insert sections on brownfields into national strategy documents, since eventual EU funding would be based on the priorities established in these strategic declarations (for example, the National Development Plan). The NGO also distributed a sample 'checklist,' organised along ministerial lines, of the kinds of amendments needed in the Czech law to reduce the barriers to brownfields reuse. This was a rough first pass, meant to elicit discussion, and illustrate the cross-cutting range of needed changes. By offering 'no strings attached' materials and consultation to those interested, the 'brownfields' topic became a vehicle through which an ambitious individual or ministerial unit could advance themselves and become more visible.

With an eye to longer term continuity beyond the life of its initial intervention, the NGO helped establish a local NGO, concerned with urban sustainability issues, which began taking on the brownfields campaign locally, finding autonomous sources of funding, and becoming the Czech partner in EU grants it helped facilitate. It began working with universities and professional organizations to increase brownfields awareness among practitioners and the next generation of practitioners, and obtained EU (LEONARDO and other) funding to prepare brownfields professional training for local authorities and professionals. And it began organizing an NGO of brownfields stakeholders, both private and public sector, to act as a lobby group and clearing house on the topic. It also began to link up with the EU professional networks on brownfields, so as to tie Central European practitioners into Europe-wide efforts.

At the risk of being banal and/or self-congratulatory, we have brought this account of NGO activity to emphasise the multiple levels at which change must occur on a topic as complex as this, and to show how the catalytic actions of a small NGO can produce fairly far-reaching systemic change within official systems. While many of the effects of this 'awareness-raising' work will take years to diffuse through the various levels of the system and translate into concrete projects, we can already see some tangible outcomes of this advocacy work in strategic documents and the Czech National Plan and guidelines for expenditure of EU Structural Funding, specific ministerial brownfields positions created (inc. at the Deputy-Minister level), brownfield funding, training, and research programs initiated, greater inter-ministerial coordination and links with the brownfields community internationally, and in concrete advice taken on several specific redevelopment projects in Prague and in several smaller towns in the Czech Republic.

The Interaction of EU Accession with the Brownfield Problem

Conducting brownfields advocacy in Central Europe during a period in which the Czech Republic was keenly engaged with accession issues, highlighted several points of mismatch between the categories and emphases of the EU accession assistance and the country's needs on an urban topic as complex as brownfields. The following are several key kinds of mismatch that hampered the EU contribution to the Central European brownfields issue. (Similar dynamics may continue to hamper EU assistance to acceding countries on other issues.)

First, is the fact that the problem has a different scale and nature than in Western European countries. There are less brownfields in these countries, they are less encumbered by bureaucracy and risks, and the market is better equipped to pick up and recycle sites whose prior use has become obsolete. For Central European cities, brownfields are of a scale and seriousness that they impinge substantially on urban competitiveness, and on the rate of greenfield conversion (with all the attendant environmental consequences). Thus, a problem that might be justifiably be less central on the EU urban agenda, is (or should be) at the centre of attention in Central Europe.

Second, is the massive horizontality of the issue, which straddles multiple institutional, temporal, and disciplinary domains, as discussed above. Urban land is not typically considered as a finite threatened resource, in the same way as water or green space. While brownfields can directly affect several media (water, soil, even air), they also entail more subtle environmental impacts (hampering the energy and infrastructural efficiency of urban forms, facilitating agricultural land conversion), as well as social and economic impacts. These are far more difficult to quantify, as are the benefits of brownfield remediation. Third, and relatedly, the range of relevant indicators and benchmarks associated with other EU priorities is far less available for the brownfield issue. Fourth, the reuse of brownfields involves the private sector almost intrinsically, in a far deeper and more sophisticated manner, than other issues. And, finally, the EU is encumbered in relating to a topic that is embedded in planning issues and institutions, toward which the EU subsidiarity principle applies.

This mismatch of emphases and categories is reflected in the limited applicability of many current EU funding programs (INTERREG, FP6) for brownfield work, and in limited EU assistance for Central European capacity-building on the topic. To remedy this mismatch, urban land needs to be included as a category of environmental protection priorities; spatial issues need to be paid greater attention, especially in a linked-up manner, in which economy, jobs, spatial form, and environment, can be jointly considered; and the range of comparative uniform indicators and monitoring systems must be expanded to include factors of relevance for brownfields

As opposed to other spheres (banking, environment, security), which received very focused technical assistance and capacity building from the EU, Structural Funding assistance on the brownfields issue seemed to primarily take the form of 'pilot projects.' The hope was, it seems, that these would build up a greater understanding by 'learning through doing,' which would lead to an increased demand for more comprehensive national tools, regulations, frameworks, programs and strategies. It is still an open question to what extent the pilot project model has been able to lead the kind of overall systemic change hoped for.

'Šternberk Case Study'

The story of one Czech town, Šternberk, shows how with some training, and lots of proactive initiative, even a very small local authority, with substantial brownfield problems and not a particularly promising location, was able to help itself tackle the brownfields issue. This small forgotten town once known for its tradition in watch making is emerging as a model of proactive local brownfield revitalization. In a short time, Šternberk became one of the first municipalities to analyse its brownfield situation, forge a partnership with the owners of one of the town's largest brownfield sites, and apply for a PHARE grant to redevelop the site for industry and housing.

With extensive brownfields, some of them heavily polluted, Šternberk was highly motivated to explore brownfield redevelopment, and in 2001 sent a representative to the ITDP brownfield seminar, co-organised by the Region of Olomouc. Over the course of three more seminars, it cooperated in preparing and testing the first version of ITDP's Brownfield Audit Project in cooperation with the Union of the Central Moravian Communities.

The audit highlighted several key findings, including the fact that seven percent of the town's total area (63% of its former industrial holding) was brownfield. The audit allowed Šternberk officials to identify a key prospective site for redevelopment and made clear that in order to achieve its objectives of providing jobs for its citizens and improving its environment, the local authorities would have to work very closely with various local private owners.

In analysing its own real estate ownership, the town realised that it had missed the boat on privatisation in allowing private owners to acquire large portions of strategically located land. To act now to reuse this land they would have to work indirectly, in partnership with these land owners.

Most importantly, the town drew on ITDP training to identify a large brownfield site that was still covered by the state National Property Fund environmental guarantees for cleanup of privatised properties. This already fragmented site was in danger of become even further broken up because of the imminent bankruptcy, which would force a piecemeal hurried sale of the remaining property. Not only would bankruptcy make future consolidation and development of the site almost impossible, it would probably strip the property of this precious government guarantee. This would have made the site truly intractable.

Thus, bankruptcy would lead to a deeper and entrenched dereliction of the property, and the resulting degradation of surrounding properties, including some owned by the local authority. Realizing the urgency and implications of the situation, the local authority approached the private owners of the site. The seminar had emphasised that a risk assessment was a prerequisite for eligibility for the environmental guarantee; since the site owners had no money to do this, the town agreed to step in and provide the risk assessment.

This assessment enabled the property owners to enter into an Environmental Clearance Contract with the National Property Fund, and in March of this year the government approved money for site clearance – a value of 63 million crowns (almost $2 million). With the first prospect of a cleanup, property regained a measure of commercial value, which will probably enable the owners to raise enough to buy back their bad debts at substantially reduced rates. The Contract will not only remove environmental pollution and ensure that site ownership remains relatively consolidated, but it established a culture of partnership between the town and the owners.

The town then drew on some of the models of creative partnerships between local authorities and the private sector presented in the workshops. In consultation with the owners, the town created a site development plan, and the city purchased some of the private property needed for infrastructure to serve the site.

The city has now applied for EU PHARE funding to help to redevelop this former industrial site as a mixed-use site for industry and housing. It has sold some of the site adjoining the brownfield to a German supermarket chain and a private businessman, in exchange for 200,000 Euro which will be used co-finance the PHARE project (co-financing is a condition for these EU projects) and a commitment from the buyers to install their own infrastructure.

The story is far from over. The site represents only one third of Šternberk's brownfields, and even if this is cleaned up and prepared it remains to be seen whether the market will take it up. But the town is well aware of the hurdles, and is building on its experience of partnership with the site owners to approach local entrepreneurs to develop their businesses in a way that could eventually come to occupy the site. In this way, an area that was once a burden on the town landscape might become the heart of its regeneration.

Figure 16.5   Picture of the Šternberk brownfields

Index

.